Photos courtesy of:
Märztagung: Marianne Tinguely, Reto Casanova
Zurich Lectures Series: Ann Chia-Yi Li, Murray Stein
Router Program, Georgia: John Hill

Translators: Irene Berkenbusch, Maria Bernasconi, Vreni Bollag, Paul Brutsche, Katharina Casanova, Andrew Fellows, Lucienne Marguerat, Urs Mehlin, Isabelle Meier, Rachel Osterwalder, Erhard Trittibach, Ursula Ulmer, Stacy Wirth, Ursula Wirtz

© 2015 ISAPZURICH with the authors
International School of Analytical Psychology Zurich
AGAP Post-Graduate Jungian Training
All rights reserved

ISAPZURICH: A Journey. The International School of Analytical Psychology Zurich, 2004-2014, Revised Edition (German edition: *ISAPZURICH: Unterwegs. Das Internationale Seminar für Analytische Psychologie, Zürich, 2004-2014*)
Lulu Press, Inc., North Carolina, ISAPZURICH, 2015
ISBN 978-1-291-96881-1

ISAPZURICH: A JOURNEY

ISAPZURICH: A JOURNEY

THE INTERNATIONAL SCHOOL OF
ANALYTICAL PSYCHOLOGY ZURICH
AGAP POST-GRADUATE JUNGIAN TRAINING 2004-2014

Editors
Isabelle Meier, Paul Brutsche, Deborah Egger & Murray Stein

Contributors
Kathrin Asper, Nathalie Baratoff, Irene Berkenbusch,
Stefan Boëthius, Paul Brutsche, Katharina Casanova,
Marco Della Chiesa, Deborah Egger, Allan Guggenbühl,
Judith Harris, John Hill, Lucienne Marguerat,
Isabelle Meier, Bernhard Sartorius, Kristina Schellinksi,
Murray Stein, Ilsabe von Uslar, Stacy Wirth & Ursula Wirtz

Revised Edition

Preface
Josephine Evetts-Secker

INTERNATIONALES SEMINAR FÜR ANALYTISCHE PSYCHOLOGIE
INTERNATIONAL SCHOOL OF ANALYTICAL PSYCHOLOGY
AGAP POST-GRADUATE JUNGIAN TRAINING

ACKNOWLEDGMENTS

First and foremost, we thank Deborah Egger who, as AGAP Honorary Secretary, brought forth and boldly pursued the idea that AGAP could found its own Zurich-based training program. We are grateful to the AGAP members worldwide who approved the delegation of AGAP's IAAP training rights in 2004 and to the AGAP Executive Committee whose members have continued to support our endeavor. Credit is due especially to the first fully constituted Executive Committee whose members worked so diligently to launch the fledgling training program in 2004: Deborah Egger (President), Stefan Boëthius (Treasurer), Katharina Casanova, Mario Castello, Diane Cousineau (Vice President), John Desteian, Dariane Pictet, Craig Stephenson, Constance Steiner-Blake, and Stacy Wirth (Secretary).

The courageous pioneers of the first generation who constituted the leadership of ISAPZURICH at the beginning were Paul Brutsche (President), Stacy Wirth (Vice President), Stefan Boëthius (Treasurer), Katharina Casanova (Director of Studies), Doris Lier (Director of Admissions), Nathalie Baratoff (Director of Program), and Karen Evers (Director of Administration). Our gratitude extends as well to other colleagues in this generation and succeeding ones named in the Appendix, who led our efforts to put down roots in the world and to thrive in this first decade.

And last but not least, we would like to thank our colleagues for the translations into either English or German: Irene Berkenbusch, Maria Bernasconi, Vreni Bollag, Paul Brutsche, Katharina Casanova, Andrew Fellows, Lucienne Marguerat, Urs Mehlin, Isabelle Meier, Rachel Osterwalder, Erhard Trittibach, Ursula Ulmer, Stacy Wirth, and Ursula Wirtz.

On behalf of ISAPZURICH
Isabelle Meier & Marco Della Chiesa (Co-Presidents)

DEDICATION

This book is dedicated to the dear ones departed since the founding: Ian Baker, Helmut Barz, Ellynor Barz, Mario Jacoby, Freya Bleibler, and to all those brave students who first enrolled in our new school.

GLOSSARY OF ABBREVIATIONS

AGAP

Association of Graduate Analytical Psychologists

AGAP is an international, Swiss-domiciled professional association of analytical psychologists and a founding member of the International Association for Analytical Psychology (IAAP). AGAP was founded in 1954 by early graduates of the C.G. Jung Institute Zurich (CGJI-ZH) with the purpose of furthering the development of analytical psychology and the professional work of its members. AGAP has grown and developed during its 54 years of existence; today it counts more than 500 members worldwide.

As an IAAP-recognized training group, AGAP has historically exercised its training privilege through CGJI-ZH. The privilege is now additionally extended through ISAPZURICH, which was founded by AGAP in 2004. Since 2006, AGAP has counted a growing number of ISAPZURICH graduates among its members. AGAP members are automatically recognized by the IAAP as analysts trained in accordance with IAAP standards. English and German are AGAP's official languages.

(Adapted from www.agap.info/active/en-/home. html, July 14, 2014)

Charta

Schweizer Charta für Psychotherapie (Swiss Charta for Psychotherapy)

Charta "... is an umbrella organization for psychotherapy training institutions and professional associations. From 1989 to 1991 a conference of authoritative training institutions [...] under the rubric of the 'Swiss Charta for Psychotherapy,' developed consensus about contents, training, scholarship, and ethics related to the essential nature of psychology. The methods represented were depth psychology, humanistic psychology, body therapy, art therapy, and expressive therapy. In 1993, 27 training institutions signed the Charta. Since that time, the agreements have been further developed and implemented by democratic process." (Our translation, from www.psychotherapiecharta.ch, July 14, 2014) Of the groups mentioned in this volume, CGJI-ZH, ISAPZURICH, and SGAP are Charta members.

CGJI-ZH
C.G. Jung Institute, Zürich (Küsnacht)

On April 24th, 1948, the C.G. Jung Institute Zürich was founded as a nonprofit, charitable foundation. The Curatorium, which administers the Institute and serves as the foundation's board of trustees is, in accordance with Swiss foundation law, self-appointing and responsible for all business. CGJI-ZH offers a variety of training and continuing education programs, all based on analytical psychology, with instruction in English and German. (Adapted from www.junginstitut.ch/english/, July 14, 2014)

Historically to date, CGJI-ZH has opted out of IAAP Membership but has required its regular faculty to be members of IAAP groups, either AGAP or SGAP. On this basis CGJI-ZH was the traditional "training home" of AGAP and SGAP, and both are still the associations through which qualified graduates of CGJI-ZH can obtain professional membership and recognition from the IAAP.

IAAP
International Association for Analytical Psychology

The IAAP, domiciled in Switzerland, was founded in 1955 by a group of psychoanalysts to sustain and promote the work of C.G. Jung. Today the IAAP recognizes 58 groups and societies throughout the world, and over 3000 analysts have been trained in accordance with standards established by the Association. The Association and its members support triennial congresses as well as other academic and clinical meetings that advance research into depth psychology. (Source: www.iaap.org, July 14, 2014)

ISAPZURICH
International School of Analytical Psychology Zürich
AGAP Post-Graduate Jungian Training

Founded by AGAP in 2004, ISAPZURICH is governed democratically with the analysts voting annually to elect the leadership and working committees, to approve the budget and the promotion of training status for colleagues, and to determine the revision of regulations, rules, and programs.

ISAP has retained the classical "immersion" model of training by which the majority of AGAP members have been trained. Today ISAP is the only

IAAP-recognized training institute in the world whose composite program of lectures, seminars, training analyses, and supervision takes place on-site in the course of two 14-week semesters each year. Within this context, ISAP offers a variety of professional and continuing education programs, all based on C.G. Jung's analytical psychology and conducted in English and German. (See www.isapzurich.com.)

SGAP

Schweizerische Gesellschaft für Analytische Psychologie (Swiss Association for Analytical Psychology)

SGAP is a group member of the IAAP and recognized as a training institute. It is "...the association of psychotherapists in Switzerland who practice according to C.G. Jung. The association has existed since 1957 and currently numbers 200 members. The prerequisite for membership is training completed at any IAAP-recognized training institute. Regular members practice in Switzerland." (Our translation, from www.sgap.ch)

TABLE OF CONTENTS

Acknowledgments ii
Dedication iii
Glossary of Abbreviations iv
Preface
   Josephine Evetts-Secker: Vita Nuovo 1
Introduction 3

WHAT LED TO ISAPZURICH'S CREATION?

Deborah Egger & Stacy Wirth: The Emergence of ISAPZURICH 7
Paul Brutsche: Opening Speech ISAPZURICH, October 23, 2004 33
Deborah Egger: Remarks on ISAP's Opening, October 23, 2004 41

OUR IDENTITY

Murray Stein: Why ISAPZURICH is Unique in All the World 47
Judith Harris: Why Democracy? 53
Ursula Ulmer: Voices from ISAPZURICH's Students 57
Stefan Boëthius & Isabelle Meier: A Profile of Jungian Analysis ... 63
Bernhard Sartorius: The Role of ISAPZURICH in Today's Society   71
Nathalie Baratoff: Ten Years of the ISAPZURICH Program
   Committee 75
Isabelle Meier Interviews Paul Brutsche: When ISAPZURICH was
   Established: The Presidency of Paul Brutsche, 2004–2008 83
Isabelle Meier Interviews Murray Stein: Steering ISAP onto the
   High Seas: The Presidency of Murray Stein, 2008–2012 97
Marco Della Chiesa & Isabelle Meier: The Classical Design of the
   ISAP Ship: Our Co-Presidency, 2012 to Date 107

OUR IMAGE IN THE WORLD

John Hill: The Jungian Odyssey, 2006–2012 115
Ursula Wirtz: The Jungian Odyssey, 2012 to Date...................... 125
Murray Stein: The Zurich Lecture Series, 2009 to Date 131
Lucienne Marguerat & Bille von Uslar: Origin and History
   of the March Conferences "Märztagungen" 137
Murray Stein: Performance Theater from ISAPZURICH 143
ISAP's Outreach to Eastern Europe and Beyond
   Kathrin Asper: Jungian Pioneers in Lithuania 159
   Irene Berkenbusch-Erbe: Renaissance for Jung in Poland 162
   John Hill: Jung in Georgia 164
   Kristina Schellinski: Every Person Counts and *Laima* Lights the
      Way 168
   Ursula Wirtz: Czech Republic, Lithuania and Estonia 171
   Bernhard Sartorius: Jung in Islamic Cultures 173
   Allan Guggenbühl: What Role Does ISAPZURICH
      Play in the Public Sphere? 177

APPENDIX
ISAPZURICH Leadership 2004–2014 183
Sources – "Isabelle Meier Interviews Paul Brutsche" 187

CONTRIBUTORS 215

PREFACE

Vita Nuovo

It is with great satisfaction that AGAP applauds ISAP's ten years of life. In 2004 this international school was born out of years of distress and discord, an experience of *nigredo* for many. It has weathered those storms and created a challenging new training institute, affirming those values first incorporated for the training of Jungian analysts in 1948 when the C.G. Jung Institute was founded. Because that year of ISAP's founding was so troubled, AGAP, ISAP's parent association, did not celebrate its half century of life. But now, as ISAP celebrates its 10th year, AGAP is in the right mood to commemorate its own 60th anniversary. The alchemists wisely taught that out of dissolution come new possibilities. So while still mourning much that was lost and honoring much that was suffered, we rejoice in the vitality and resourcefulness of this training program. The analysts who brought it into being were courageous and shared a vision of faithfulness to core Jungian values that must not be abandoned in an *aggiornamento*. Some of those highly esteemed colleagues have not lived to observe this decade; we remember and pay tribute to them in our festivities: Ian Baker, Ellynor Barz, Helmut Barz, Freya Bleibler, and Mario Jacoby. We owe thanks to many who helped create ISAPZURICH, and those who continue to serve it, generously giving their time, energy, and commitment. The new premises at Stampfenbachstrasse are tall, light, and airy; coming here for the first time felt like emerging from cellar to lighthouse. Leaving that first home at Hochstrasse was hard though; it had welcomed and sustained ISAP's first tentative beginnings. But now a confident school moves forward with vigor and integrity and AGAP salutes you and wishes you many more decades of commitment to Jungian psychology.

Josephine Evetts-Secker, 2014
AGAP President

INTRODUCTION

ISAPZURICH has been underway for ten years and has travelled a considerable distance. After a decade, it is time to pause and reflect on the road taken, which was sometimes hidden sometimes in the open, expanding as if alongside a river, whose bubbling and murmuring sounds constantly accompanied those travelling along its shores.

We look back to a beginning that was etched in pain and grief, anger and disappointment at the state of affairs at the C.G. Jung Institute in Küsnacht in 2004. The initial stage of departure was highly emotional and this energy constellated into the determination to go our own way under the auspices of AGAP. Without the fuel of these emotions, the separation and establishment of a new identity would not have been possible. As sad and upsetting as the departure was, this emotional charge gave ISAPZURICH the chance to set out on its own pilgrimage and thereby initiate a process of individuation.

This new body of analysts wanted to participate in a strong democratic process, and thus began to develop a formal, structural variation of the old C.G. Jung Institute in Küsnacht. In retrospect, we are deeply grateful for this democratic structure and remain committed to it. ISAPZURICH originated from the strong bond of Eros nurtured among the many founding analysts, and it continues in this spirit. Analysts want to be a part of the decision-making process as they work hard to sustain ISAP's future. It is challenging to work in this way and not lose touch with spirit and Eros, but the rewards are even greater when this combination succeeds.

Substantive differences in the training programs now also distinguish us from the C.G. Jung Institute in Küsnacht. Most recently, ISAP's firm decision to maintain its classic form of training has renewed and strengthened its vision for the future. ISAP is not the only Jungian training program that wants to retain this classic form. We share this perspective with the C.G. Jung Institute of Berlin and the Society of Analytical Psychology's Institute in London. This position stands in stark contrast to other institutes that have adapted their educational and training standards to align themselves with the demands of the state. Thus, ISAP continues its journey of individuation with its original intention and commitment to the classic form of Jungian training.

We have now arrived at a plateau from which to look at the path travelled in these first ten years—a century-old post office building from the Art Nouveau period. In this anniversary publication, we review the road travelled and talk about our accomplishments and our identity. What the future may bring is open and should remain open. But one thing is certain: The community of analysts at ISAPZURICH will continue accompanied by the spirit of Eros that brings life and instills dedication to share Jungian psychology with the world.

The Editors: Isabelle Meier, Paul Brutsche, Deborah Egger, & Murray Stein

WHAT LED TO ISAPZURICH'S CREATION?

THE EMERGENCE OF ISAPZURICH

DEBORAH EGGER & STACY WIRTH

Many threads weave a fabric. Many streams feed a river. Many parts make a whole. Whichever metaphor one uses, one part alone cannot determine the end result. Neither can the whole be imagined from the perspective of the parts. So it was with the emergence of ISAPZURICH. Over the years that this one brief chapter can only begin to cover, the streams of events and experiences that converged into what we now celebrate as an established ISAPZURICH were arising independently, but not, as it turns out, without meaningful connection.

How is it possible that a large core of dedicated and faithful teachers, trainers, and leaders of the C.G. Jung Institute Zürich could end up turning against the very place they called "home"? How is it possible that the same people who had been keepers of the long and vital tradition of the one and only Zurich Institute could even begin to imagine leaving it? The idea of something like ISAPZURICH was not even a remote glimmer of possibility in the beginning. What happened in the hearts and souls of individuals to change them so profoundly that they had to do the unthinkable, the unimaginable? This intense, mysterious, and almost miraculous story is the focus of this first chapter: The Emergence of ISAPZURICH.

A Question of Loyalty: Finding One's True North

From the time of the Zürich Institute's founding in 1948, the analysts conducting training had done so freely, bound by good faith in the leadership; that is, the Curatorium, which is self-electing in accordance with Swiss foundation laws. Because the analysts do not elect the Curatorium and have no binding say in institutional matters their loyalty to the Institute is all the more meaningful. The very question of loyalty was often at the center of discourse between 1997 and 2004. Emotions on the topic ran high in the community from the beginning of the disclosure of conflict within the Curatorium.

On April 1, 1997 the Curatorium elected Brigitte Spillmann as president. Paul Brutsche, her predecessor, had served five years and during his tenure had developed a spirit of cooperation and engagement among the analysts with his warm and congenial style of leadership. In a continuing gesture of his democratic spirit, he had led the Curatorium, encouraging the community of analysts to recommend candidates for his replacement. When junior analyst Brigitte Spillmann was nominated to succeed him, the Curatorium accepted her wholeheartedly. It did not take long, however, for conflict to arise. Most of the early strife appeared to radiate from President Spillmann's leadership style and later also proved to be based on her troubling initiatives and plans (that, in the view of many analysts, ended up being destructive for the Institute as well as for the community). Decades of friendships and collegial bonds were torn asunder as individuals found themselves tossed and catapulted from one series of events to another.

As struggles arose between and within groups, colleagues were faced with self-doubt, accusations, projections, wounded feelings, and mandates—all circling around feelings of loyalty to self, others, and the community. Regardless of the side taken in any given circumstance, each of us was required to manage our loyalties, to find our *true north*. The coagulation of the (ultimately cohesive) group of analysts that ended up founding ISAPZURICH was a bumpy process, filled with endless discussion, debate, and agony. Each individual had to search his or her own inner territory to be able to continue every step of the way. Whichever side of any debate one fell on was ultimately a personal decision that had to be respected. The task, the trip, was arduous.

The emergence of ISAPZURICH is rooted in the loyalty of generations of analysts trained in Zürich and faced with this turmoil. In the early months the ferment was contained within the Curatorium, but it quickly impinged upon the Institute's body of analysts. On September 8, 1998 in the events hall at the Weisser Wind, the Curatorium majority (Paul Brutsche, Helmut Barz, Kathrin Asper, and Elisabeth Hartung) requested that Brigitte Spillmann step down from the presidency as a result of the discord surrounding her autocratic style of leadership. They suggested that she remain on the Curatorium, but not as President. The Weisser Wind meeting had been arranged by a group of analysts supporting Brigitte Spillmann. Pressure to take sides in a Curatorium-splitting dynamic became palpable at this historic meeting.

At this extraordinary gathering, the large majority of the 100+ mostly uninformed analysts supported Brigitte Spillmann's desire to stay in office and continue her mandate of reform. The Curatorium majority felt they could not go against the will of the vast majority of analysts that evening, even though they had the statuary power to do so. Instead, they decided to honor the community's wish and try to work out their differences. The day after the Weisser Wind meeting, the Curatorium pled for the community's "patience, trust, and cooperation."[1] But things did not improve. In the ensuing months and years, the community witnessed seemingly innumerable contentious meetings, resignations, firings, and restructurings of committees and operations, only some of which you will read about here.

The disturbing conflicts exposed at Weisser Wind led a small group of analysts to meet regularly and informally, going under the rubric of SOS, an important initiative from Diane Cousineau. Little by little, other colleagues found their way to these meetings, which encompassed exchanges of personal experiences, psychological reflections on the situation, and always an Eros-filled sharing of meals.[2] These analysts found themselves conflicted, wondering how to stand faithful to an Institute whose identity and values they felt were being transmogrified.

At the same time, some of these analysts continued to work on the Institute's fundraising committee, which had been established under Paul Brutsche's presidency (1992-1997) and consisted of an English-speaking group of colleagues. They launched an annual publication titled *The Bridge*,[3] whose twofold purpose was to solicit alumni and friends for financial donations to the Institute and to share news of the Institute with the outside world. *The Bridge* editors soon sought the support of Joanne Wieland-Burston, AGAP's Honorary Secretary and cooperation was forged between AGAP[4] and *The Bridge*. *The Bridge* quickly became successful funneling donations and bequeathments to the Institute. The second issue published feedback from Hans-Rudolf Kuhn, who was then the Business Administrator of the Institute:

> The first Bridge provoked many responses, a very special one I want to share with you now. It came in the form of a letter from an analyst who wishes not to be named. "I'm writing this letter on the occasion of reading the Bridge," the analyst wrote, informing us about having set up a will in which the Institute is a beneficiary. We would like to express our appreciation and thanks for this most welcome generosity.[5]

Given the loyalty devoted to a project like *The Bridge*, the Curatorium's dismantling of the fundraising committee in the fall of 1999 met with the committee's disbelief and disappointment. In a fateful meeting, John Granrose, Director of Studies, came to inform the committee members that they no longer needed to meet. Deborah Egger, chief editor of *The Bridge*, asked, "Have we been fired?" Mr. Granrose answered, "Yes."

One after the other, committees that had been established during Paul Brutsche's presidency were dismantled under the auspices of the Institute's restructuring, which was begun in 1999 and orchestrated by Mundi Consulting AG, brought in at Brigitte Spillmann's initiative. This restructuring was presented as a response to new professional demands as well as to the need to modernize the Institute's infrastructure. But in the course of reform, one could feel a growing divide between the languages and cultures at the Institute, something particularly difficult for the multi-lingual and multi-national community and its loyalties. As it evolved, this subtle shift seemed to be less about modernization and new professional demands. Neither was it *cleanly* delineated along language or cultural lines, but more ominously based in complex territorial power issues, which eroded the cohesion of the community. Analysts involved in the Swiss-oriented, German-speaking, child and adolescent program found themselves poignantly situated at this crossroads. This specialty area was very much encapsulated within the local context, so much so that many child analysts felt no professional (or personal) option other than to accept all that was happening at the Institute. Finding one's true north was not easy, not easy at all.

Growing Crisis: Leadership, Staff, Training, Finances

As the crisis at the Institute escalated, so grew the number of analysts meeting informally as the SOS group. With the increasing intensity of the conflict, the meaning of SOS—"Save our Souls"—gained gravitas.[6] At the beginning no one could have known how far the conflicts would carry us. Looking back we can now discern how the mounting crisis impacted four crucial areas in inseparable ways: *leadership, personnel, training focus* and *fiscal management*. The fact that so many areas of Institute life were affected simultaneously did a lot to swell the streams of discontent, alarm, and need for change.

An example of the cause for concern with the leadership can be seen in what happened with the analysts' consultative votes of 1998 and 2000:

At the resignation in 1998 of three Curatorium members (Kathrin Asper, Helmut Barz, and Elisabeth Hartung), the four remaining members (Paul Brutsche, Christina Hefti-Kraus, Robert Hinshaw, and President Spillmann) accepted the analysts' consultative vote for three new members: Dorothee Imboden-Lanz, Andreas Schweizer, and Ernst Spengler. The Curatorium also obligated itself to fulfill the "explicit provision" that in two years a second consultative vote by the analysts for the renewal of the Curatorium would take place and be honored.[7]

The next consultative vote took place on March 24, 2000, as promised. But without publishing the counts, a Curatorium majority rejected the results, justifying their action with the perspective that "the way the discussion developed and the final outcome seemed incongruent and there was hardly any debate about substantive matters." They went on:

> ... we have the result of a meeting of analysts, which can only be consultative in nature since such a meeting is not mentioned in the statutes governing the Institute. No meeting of analysts can relieve the Curatorium of this duty.[8]

Five days later Curatorium member Paul Brutsche independently revealed that the election results, in fact, would have removed Brigitte Spillmann from the seven-member Curatorium. The results of 100 votes were,

> Andreas Schweizer (incumbent): 77; Ernst Spengler (incumbent): 71; Paul Brutsche (incumbent): 71; Dirk Evers (new): 70; Francoise O'Kane (new): 66; Daniel Baumann [new]: 61; Dorothee Imboden-Lanz (incumbent): 60; Doris Lier (new): 52; Brigitte Spillman (incumbent): 48.[9]

In addition to ignoring the election outcome that would have squarely placed Dirk Evers in the Curatorium and clearly eliminated Brigitte Spillmann, the Curatorium majority decided that Brigitte Spillmann would remain President. As she explained to the Swiss society SGAP at their assembly in May 2000,

> It saddens and burdens me very much that on 24 March I received only 48 votes and nevertheless stand before you today as president of the Curatorium ... On that evening I departed ... with painful consciousness that the consequence of this result could only mean my immediate withdrawal from the Curatorium ... There followed discussions with colleagues of the Curatorium ... The effects of this withdrawal in the current phase of restructuring were judged to be irresponsible and destructive, and this especially because no factual arguments for the vote of mistrust were put forward, and because no acceptable alternative presented itself, and because beyond this my withdrawal would have brought others. Thus it finally came to the known decision of the Curatorium majority to adhere by the current [Curatorium] constellation and my presidency.[10]

The next round of changes to the Curatorium happened merely five months later in August 2000 when Paul Brutsche, Dorothee Imboden-Lanz, and Andreas Schweizer resigned and were replaced by Ruth Ammann, Felix Huber, and Irene Lüscher. An unprecedented parade of members marching through the Curatorium continued over the ensuing years. At the same time, the President and those Curatorium members who supported her persisted in arbitrarily applying rules—which was done not only with regard to the foregoing consultative elections, but in other ways as well. This seemingly callous use of power alienated increasing numbers of analysts who experienced this development in many settings, including the way agendas were set, meetings were run, printed material was distributed, jobs and roles were dismissed and reassigned. This sense of power abuse mightily fed the river of change, no less in the way it affected long-standing and loyal staff.

Within one year, from March 1999 to April 2000, the Jung Institute lost *four* valuable staff members. Their resignations or firings were accompanied by their identical expressions of dismay at the worsening work atmosphere. The perhaps well-intended restructuring of the Institute contributed, sadly, to increasing unhappiness and unease. Under a stringent top-down management style, employees lost contractual benefits and felt over-controlled and subjected to ungrounded criticism. A climate of anxiety developed following absurd limitations on personal conversations, the imposition of rigid lines of authority, and the demand for absolute cooperation and loyalty with threats of consequences for disobedience. A one-sided emphasis on efficiency overran past practices that had nurtured dignity and respect.

For many of us these experiences echoed the very definition of poor leadership style, which, based on mistrust, strives for control by the use of intimidation, degradation, and micromanagement. This contrasted an earlier positive leadership style that had been based on trust and supported individual creative work, responsible engagement for the whole endeavor, and vital self-esteem. Reading the brave, but heartbreaking, letters of dedicated staff members who were employed at the Institute at that time brings home the magnitude of the loss to all involved: personnel, students, and analysts.

Strangely, by February of 2000, Mundi Consulting AG saw the "mutation in the office and turnover in the personnel" differently. While acknowledging the intense challenge placed on the staff, consultant Ueli Mühlemann assessed, "Already within a short period of time the climate between the management of the Institute and its staff has improved noticeably. The insecurity, still tangible early last summer, seems to have given way to a certain optimism."[11]

To the contrary, when one reads the letters and other documentation from various staff members during this period, one finds descriptions of experiences like "the dismantling of my position and competences," "the impression that everything the staff has done up till now is wrong," "my work is said to be full of mistakes and my motivation questionable," "I should learn how to cut off a conversation and be more efficient rather than friendly," "the analysts have no say in the matter, I should not believe they stand behind me, I am subordinate to the Curatorium"—which go on and on. The discrepancy between what people were experiencing and what was published as the official situation is flabbergasting and far from the alleged "optimism." Such contradictions consistently contributed to the corrosive splitting dynamic emanating in messages from the Curatorium to the community.

The stream that carried the changes of focus in training was more subtle and stemmed from the evolving Swiss political scene and regulation of psychotherapy. The political acuity with which the Curatorium approached the changes in Swiss law certainly reflected the Zeitgeist. On the other hand, it resulted in less energy flowing into the *international* analytic training program. This shift in focus was compounded by the above-mentioned growing, corrosive splitting dynamic.

Intuitive foresight of this subtle shift fed a sense of urgency within the SOS to save the international analytic training. This quiet, but essential, stream has been definitive in the evolution of ISAPZURICH's training identity. We see now that the international training at the Institute is conducted in three 3-week blocks with training analysis no longer Zürich-based and supervision not contiguous with casework. New Swiss federal law now requires psychotherapy training, among other things, to conduct a mere 50 units of individual analysis or therapy within one's own training program—this amount is the prerequisite just for *applying* to training at ISAP![12] There is a wide gap between such modes of training and ISAPZURICH's immersion models.

The current of financial change was felt most acutely. Repeated arguments about the looming financial crisis began in 2001 with a report of the Institute's first serious deficit of CHF 126,700.[13] The Curatorium treasurer at that time, Daniel Baumann, assured the community:

> The Institute possesses approximately Fr. 3,500,000 useable capital and thus has a very solid financial basis. ... The consequences of past conflicts and differences among the analysts have also affected both the income and the expenditures of the Institute: the dramatic reduction in donations to the Institute is more than symbolic. ... In the mid-term (about three years) the administrative expenditures of the Institute should be covered by the income generated. ... The private capital of the Institute (about Fr. 1,900,000) was only reduced by Fr. 126,700.[14]

In the course of 2001, patron and former Curatorium Treasurer Antoinette Baker, along with colleague and former financial advisor to the Curatorium Hans-Peter Kuhn, drew the Curatorium's attention to the fact that, as compared with the current loss of CHF 126,700, within two years, operating losses of CHF 321,160 had contributed to a total loss of an astounding CHF 433,300.[15] Indeed several analysts with financial expertise repeatedly warned the Curatorium about overexpenditure, especially in view of the declining student enrollment since 1998. Such warnings were consistently rebuffed, and this kind of ping-pong dispute went on until 2003 when the Curatorium finally admitted to a real fiscal crisis.

In meetings on January 17 and February 7, 2003, the Curatorium revealed that the operating capital of the Institute was in rapid decline, attributing the cause to a combination of factors. Besides decreasing student enrollment, these included poor stock market performance, the start-up of an ambulatorium (Care Point), and the expansion of the premises to Theodor-Brunner-Weg. Emphasizing "the financial straits" that put a "decades-old tradition ... at stake," and that "time [was] of the essence," the Curatorium announced measures to decrease expenditures and increase revenue; in the short-term, the needed revenue was to come from the analysts' time-limited "cost-sharing" in defined amounts.[16] The majority of analysts were willing to support the institute financially. However, at this point, they would do so only under the following conditions: that,

> (1) ... reflection on the reasons for the financial crisis and the crisis of trust at the Institute would take place; (2) an expert commission with the power of co-decision would be established and have access to and monitoring of the Curatorium's financial strategy; and (3) the Curatorium would, as a measure of trust, subject itself to election by the analysts and ratify the results.[17]

The Curatorium refused the conditions and the financial crisis intensified in March 2003 at an SGAP assembly. Voting on spontaneous motions from the floor, the vast majority of analysts resolved to (1) create an anonymous donation fund for SGAP by which individuals could support the Institute and simultaneously protect their privacy and (2) establish a commission of experts to evaluate the Institute's financial situation. However, because these approved motions were not on the assembly agenda and were therefore non-binding, SGAP members Irene Lüscher, Brigitte Spillmann, Felix Huber, and Ernst Spengler—who were also Curatorium members—took the opportunity to respond with legal action. At considerable legal expense to SGAP, a settlement with the Justice of the Peace annulled both resolutions.[18] The whole affair failed to improve the Institute's financial situation or relations between the Curatorium and the majority of analysts.

Incredulous at the astounding lack of possibility for dialogue and movement and in an attempt to break the stalemate, colleague Stefan Boëthius launched an independent email forum, inviting all analysts of the Institute to share their opinions, ideas, and feelings about the situation. This afforded honest exchange, sometimes heated debate, and above all leveled the communication field, providing participants with an equal voice and sense of agency.

Nevertheless, the crisis escalated massively in May 2003, when the Curatorium declared that the analysts' financial contributions were not voluntary donations, but mandatory and *ad personam*. By mid-June, 144 analysts had voluntarily entered their names on a list of non-payers, which was sent to the Curatorium with a reminder of the conditions under which these analysts *would* be willing to pay. One month later the mandatory contribution was coupled with the requirement to sign a Declaration of Affiliation by which one was to declare one's readiness to pay contributions in undesignated amounts and for an undesignated duration of time. Those analysts who did not sign by September 1st would be "stricken" from the Catalogue of Analysts, meaning that they would immediately lose their functions at the Institute.[19] This move on the Curatorium's part incensed so many analysts that the 144 "non-payers" then created what would become a rapidly growing list of "non-signers." This is only one example of how the Curatorium's ever-tightening vice-grip served, paradoxically, to stoke the strength and vitality of the current against them and favor change.

In mid-September 2003 some 90 deeply distressed analysts gathered in the meeting hall at Erlengut to discuss what alternative action could be taken.[20] All agreed that the Curatorium had no right to demand financial contributions and this requirement would not be entertained.[21] Moreover, the majority wanted the Institute statutes to embrace a more democratic form of governance. They also agreed to the option of withdrawing from teaching if the Curatorium did not enter binding negotiations with the SGAP Board, who had agreed to represent the analysts' interests. Finally, because the students lacked concrete information and, in the view of a large group of analysts, were fed disinformation and so suffered from ominous rumors the analysts decided to inform them about the outcome of this meeting.

In accordance with the wishes of the Erlengut participants, a letter from the Erlengut Coordination Committee (dated September 15, 2003) apprised students of the Erlengut developments.[22] Shortly afterward the same committee invited the students to a November meeting with the analysts. At this event students were informed that a number of analysts would offer a "supplementary" program of lectures and seminars in January and February 2004 to be held in private offices, homes, and rented spaces, since their teaching at the Jung Institute was no longer possible, except as "guest teachers." The Curatorium promptly sent registered letters to the Erlengut committee, reprimanding them for "damaging" the Institute and threatening loss of training status for any such repeated behavior.[23]

By the end of September 2003, the list of 144 non-payers had swelled to 183 analysts who publicly declared their unwillingness to sign or make payments to indicate their "affiliation" under the prevailing conditions. The new analyst directory released on November 1st showed approximately 220 names had been stricken from the original list of 316, leaving the Institute with 100 "affiliated" analysts, approximately 55 of whom were left to teach a full English program for international students.[24] With profound disappointment a number of analysts soon learned that, in the very same month when all of this was going on, Brigitte Spillmann had apparently dismissed the "anxious concern" conveyed by IAAP President Murray Stein on behalf of past IAAP presidents and the president-elect; worse, she had ignored their offer of "round table" discussions that held out the hope to "rise above partisan interests and personal feelings and to make a dedicated effort to keep alive the flame of Jungian thought, teaching, and research in the city where Jung himself lived and worked and taught."[25]

Following the Erlengut meeting, the Curatorium did, however, agree to begin negotiating with the SGAP Board. When this process achieved no binding resolution, the Curatorium and SGAP Board entered into mediated negotiations with Professor Konrad Osterwalder, then rector of the ETH Zurich.[26] During this negotiation period in early 2004, Prof. Osterwalder established a voluntary and anonymous support fund for the Institute. The Curatorium, in turn, agreed to temporarily suspend its demand for mandatory payment and to temporarily restore the Catalogue of Analysts as it existed at the end of October 2003—but which "[did] not imply restoration in earlier functions on committees and other bodies of the C.G. Jung Institute."[27]

By the end of April 2004, the SGAP Board had felt compelled to withdraw from these negotiations explaining their reasons,[28] which the Curatorium soon rebutted in a letter that claimed to state "the facts."[29] Formally closing the negotiations at the beginning of May, Prof. Osterwalder noted the SGAP's proposed structural reform, a model that aimed at constructive cooperation between SGAP and the Curatorium. These negotiations also ended with no binding resolutions, and Prof. Osterwalder reported that donations of CHF 36,510 had been collected for the Institute, an amount that fell short of expectations.[30] At the SGAP assembly on May 15, Prof. Osterwalder was asked for his opinion about the Curatorium's communications on the matter. He replied that their letter constituted "a very subjective representation. [...] But I would like to concede to you," he added, "that it is the Curatorium's interpretation and the letter was not OK'ed by me, I had not seen it before."[31]

A highly frustrated and disappointed SOS group, numbering 45 colleagues, had resolved to force a last-ditch ultimatum with the Curatorium to try to save the Institute.[32] They launched a pledge campaign, with the promised funds to be handed over to the Institute under the conditions that all Curatorium members step down at the end of the summer semester 2004 and that a new Curatorium be elected by the analysts' consultative vote. The Curatorium never replied to the May 17 notification of pledges from 169 analysts in the amount of CHF 350,444 ready to be given to the Institute when the conditions were met.

AGAP's Development and Position During the Crisis

When the SOS group met at the end of May 2004, they were stunned at the Curatorium's silence and at the realization that they were now faced with an unrelenting deadlock. Weary of the strife and desperate to work again under constructive and creative conditions, the group was ripe for what happened next. On that night AGAP President Deborah Egger sketched on a flip chart a possible configuration that had been shared with an initiative group from the SOS, which had also been working for a few months on necessary aspects of a new training program. The idea was to create a sub-group of Zürich-based AGAP members to whom AGAP's IAAP training privileges could be delegated. It was met with resounding relief and work began the same night on a concrete plan that would be presented to the AGAP Executive Committee (ExCo) and then presented for vote at the AGAP assembly in Barcelona in August—just three months away.

Since time was so short, the ExCo had to work closely with the SOS initiative group to efficiently lay out a plan of action. The invitation and documents for AGAP's Barcelona assembly, due to go out in June, had already been composed and were being translated in preparation for printing and mailing. Completely separate from the SOS initiative, AGAP already had enough big ticket items on the agenda: a proposed dues increase, and the election of new committees—a nine member (international) ExCo, an Ethics

Committee, and Auditors—as required by the new Constitution (*if* it got approved by circular vote ahead of time). And proxy voting was being introduced for the first time. On top of it all, to insure a binding vote, the new motion, "Proposed Delegation of Training to Zurich Sub-Group," had to be added to the agenda. This meeting was promising to be a blockbuster from many points of view and, indeed, it would turn out to be.

But how had all this come together at one point in time just when a new AGAP Constitution was also being considered? We need to back up and take a look at AGAP's development since 2001 to understand how and why these separate but related events converged. In August 2001, at the AGAP General Assembly in Cambridge, Joanne Wieland-Burston stepped down after her tenure of eighteen years and Deborah Egger was elected president. Several items begged attention, including the orientation of the newly elected AGAP ExCo that consisted for the first time of an "outer" and "inner" circle. As AGAP had grown so large over the years, it was impossible to imagine leading the Society without representation from the membership around the world, thus the creation of the two "circles." Also in Cambridge, AGAP was elected to sit on the IAAP Executive Committee, which gave it a stronger voice on the international stage. This was a fruitful time for the AGAP president as she learned the workings of many IAAP societies and reflected on the challenges AGAP faced with its archaic constitution, lack of structure, large membership, and global distribution.

There was a reasonable request from the IAAP that AGAP's ethics code be updated to meet the international minimum standard, which would include the creation of an AGAP Ethics Committee for the first time. In addition, AGAP's lawyer, Martin Amsler, noted that AGAP's Constitution, based on Switzerland's 1954 association law, was out of compliance with current law and had to be substantially updated. According to the outmoded constitution, this update could only be achieved with a worldwide circular vote. The ExCo began work with Mr. Amsler in 2003 to prepare this vote in advance of the next AGAP General Assembly, which would take place in 2004.

As the work progressed, the many problems in the old constitution led Mr. Amsler to recommend the adoption of a new constitution. The ExCo agreed and so, after many hours of work, a new constitution was ready for presentation; in June 2004 it was circulated for vote to all AGAP members, allowing time for several weeks of discussion and debate. The most important points of change fell into three categories: (1) structural—dealing with terms of office, committees and membership; (2) ideological—functions and goals of the Society; and (3) ethical—to bring the ethics code in line with IAAP minimal standards.

Meanwhile, most AGAP members were aware of events unfolding at the Zürich Institute and were keen to hear news of their beloved alma mater. From an AGAP viewpoint, the crisis revolved around the professional difficulties faced by its members as a result of the crisis in Zürich. For instance, even though by November 2003 approximately 100 Zürich analysts had acquiesced to the Curatorium's demand for mandatory payment, more than 200 analysts—most of whom were AGAP members—had not. The non-compliant colleagues found their training functions revoked with the exception of analysis and supervision begun before the current semester. The AGAP Executive Committee began to be concerned about the long-term impacts on the international program and the threatened livelihood of Zürich-based AGAP members.

As tension built in Zurich in late spring of 2004, it also became clear that AGAP members at large could perhaps play a major role in the shaping of events in Zurich. The AGAP Executive Committee, based in Zürich, discussed and debated ways to support members on whichever side of the conflict they found themselves, which was not an easy proposition. The ExCo also developed a solid working relationship with AGAP's sister group, the SGAP, and kept the IAAP ExCo abreast of developments. The AGAP ExCo shared this summary perspective with the membership in early 2004:

> While not all of our members are in Zurich and while many of us are involved in Institutes and Societies around the world, our historical and emotional ties to the Zurich Institute are in no way diminished by our participation elsewhere. In so far as AGAP supports the Institute financially through the steady funnel of students encouraged to study in Zurich via Alumni as

well as through the Student Loan Fund, we have a vested interest in the direction the Institute is headed.

Therefore, we wholeheartedly support a plan that not only addresses the financial crisis but also the crisis of trust, respect, and spirit that continues to plague the well being of the Institute. In addition, we affirm the continuing requests from the community of analysts to the Curatorium that there be a democratization within Institute policy and functioning, even as the Institute remains tied to its "Stiftung" (Foundation) status. We have heard and likewise affirm the willingness on the part of the analysts to contribute financially to the Institute during this crisis, provided that there be a new working relationship between the community of analysts and the Curatorium and that this be agreed upon beforehand.

We also support the re-focusing of the Institute's resources on its central training programs, which have traditionally offered a high caliber of individual analytic training in both German and English for people of all lands bringing with them the richness of a wide variety of professional backgrounds. To this end, we are aware that AGAP, like the SGAP, is a fully qualified training Society of the IAAP and stands ready to assist actively should the need arise.[33]

A well-used email forum had invited all AGAP members to discuss the proposed new constitution during the summer of 2004. Many opinions were exchanged as members around the world wanted to be sure they understood the changes and their implications. The Curatorium vowed to bring legal action against AGAP should the membership approve the new constitution, arguing that AGAP was nothing but an alumni association, with no right to conduct or delegate training; therefore, to state AGAP's training right in a new constitution would be to violate our society's founding purpose.[34]

While the AGAP ExCo had steadily informed its 600+ members around the world about the developments in Zürich, the IAAP also continued to follow the situation closely. IAAP President Murray Stein set the record straight with the Curatorium in July 2004 when he replied on behalf of the entire IAAP Executive Committee to Brigitte Spillmann:

> I was quite surprised to find the Curatorium now addressing the IAAP Executive Committee with a concern about AGAP after you declined my offer last September to form a "round table" in the interest of mediating the conflicts between the C. G. Jung Institute Zürich and the two IAAP member Groups in Switzerland. It is regrettable that something could not have been done earlier to prevent this further exacerbation of tensions and conflict...
>
> From the IAAP's point of view, AGAP is not merely an alumni association of the Zürich Institute but rather a regular member Group and as such is entitled to develop its own internal structures, Constitution and By-Laws, and eventually set up its own training program if it so chooses so long as this complies with IAAP requirements for analyst training ...
>
> As an historical note, it should be said that AGAP was kept as a member Group of the IAAP after some debate in the early years because the Jung Institute Zürich does not fall under the jurisdiction of an IAAP member Group. As a consequence, its graduates cannot enter into the IAAP in the normal way...
>
> Without AGAP, many graduates of the Zürich Institute would not be included in the IAAP membership today. From the IAAP's point of view, therefore, AGAP is an exceedingly important and valuable member Group. It seems to me, too, that it is strongly in the self-interest of the Jung Institute Zürich to maintain good relations with AGAP, since this is the IAAP member Group whereby its graduates are most easily able to enter into and become a part of the IAAP.[35]

In spite of the Curatorium's threat of legal consequences, at the end of August 2004 the AGAP members did indeed approve the new Constitution, ending the circular vote resoundingly with 392 "yes" votes to 24 "no" votes.[36] The assembly in Barcelona was now required to run on the basis of this new Constitution.

During the summer of this voting, there was plenty to do in Zürich. The AGAP President issued an invitation to all analysts working in Zurich to attend a meeting on July 6th 2004 at St. Andrew's Church. The purposes were to present a structural model for a new training program, to discuss the possibilities of what a positive vote in Barcelona would imply, and to gauge the general local interest for the proposal. Of the nearly 100 colleagues attending, the vast majority approved and began the initial work of creating committees to establish the new training program.

All summer long, the initiative committee and AGAP's ExCo worked with Zurich colleagues on the unending details involved in setting up training, including preparing for the possible start-up meeting in September. Fortunately, they were all quite experienced trainers and committee members, so the work was intense and fruitful. And there was hope. Finally there was light at the end of the tunnel! If only AGAP members at large would approve and lend their support to the crucial vote.

Under considerable tension at the AGAP assembly in Barcelona and with 149 "yes" votes and 17 "no" votes, members did approve the delegation of AGAP's IAAP training right to a sub-group of Zürich members.[37] Gripped by the enormity of the historical moment in analytical psychology that we were experiencing and participating in and feeling bittersweet relief of an unwanted battle hard won, we welcomed Murray Stein's reminder that, as a democratic association, we could revisit this vote at our next, or any future, assembly.[38]

Convergence, Emergence, and Consequences

Under the balmy Barcelona sky, a group of future ISAP colleagues sat together that evening, feeling the elation and relief as immense as the feelings of solemnity and devastation. We recounted the hundreds of hours of work that had gone in to making this opportunity a reality. We struggled to take in the magnitude of the responsibility and good faith that AGAP members had just handed to us.

After the Curatorium failed to respond to the large pledge of financial support, our last hope of remaining at the Institute was dashed, and we had worked madly to be able to go our own way. Fully exhausted, we were now faced with going back to Zürich and *really* working to pull off the opening semester of a new training curriculum within a few weeks. The first organizational meeting took place September 9, 2004 in Erlenbach, when members elected the leadership and other committees.[39] There was still a lot to take care of. At this juncture we still had no students, no physical premises to conduct training, no staff or infrastructure, not even a roll of toilette paper for the powder room we did not have. Beyond the convergence of the many streams thus far, it seemed that miracles were needed to open the doors of a new training center within six weeks.

Some amazing things fell together as word spread and connections were made. When Doris Lier told her friend Herbert Maissen that our new school needed rooms, he answered, "I have rooms." As talks progressed, Mr. Maissen's generosity and support in the form of upgrades to the building on Hochstrasse 38 and providing furniture and flexible lease terms proved crucial and definitive to ISAP's start. As it turns out, when Mr. Maissen co-founded AKAD—a private academy for adult education—a friend of his had supported the endeavor by providing affordable space, and so now Mr. Maissen felt compelled to pass this gift on to ISAP. And what an important gift it was! Also, several hundreds of thousands of francs from within and without Switzerland were given to ISAP at the start, including money for student loans, a library and general operation expenses. These timely and astounding gifts also proved fateful in securing students and the implementation of training. Step-by-step, things fell into place until a packed house of analysts, students and well-wishers gathered on October 23, 2004 to celebrate the opening semester of ISAPZURICH. More detail is given about this event in the opening speeches contained in this volume.

Just as we were feeling like we had gotten through the worst, we received notice of the expected lawsuit against AGAP, filed now however by three AGAP members: Curatorium President Brigitte Spillmann, Curatorium Member Ernst Spengler, and Robert Strubel, a colleague from the Institute. This was the third time in which then active Curatorium members or an analyst faithful to President Spillmann had brought lawsuits against colleagues— in one case, against a diploma candidate. Well into March 2005—during the legal proceedings against AGAP—the Curatorium worried ISAP students with such written claims as "... AGAP's so-called special training program does not lead to a secure and legally recognized degree."[40] Dismissing Murray Stein's reasoning of July 2004, the Curatorium insisted

Acceptance as a member of the IAAP is also questionable, since the legitimacy of this training is legally disputed within the AGAP framework. A related lawsuit contesting the decisions taken within the AGAP is currently underway and could drag out for some time. As long as this remains in legal question, one cannot assume guaranteed acceptance in the IAAP. Hence this new training program makes your career prospects highly uncertain.[41]

Not only did the AGAP ExCo and Murray Stein again set the record straight but the new IAAP President Christian Galliard wrote,

... I have the pleasure of confirming that AGAP is one of IAAP's Constituent Societies, designated as such at the Constitutional meeting held in 1956. It is a recognized Group Member of the IAAP with training privileges. ... AGAP exercises this training privilege through the International School for Analytical Psychology (ISAP), and ... upon successful completion of the ISAP training and receipt of the diploma, a graduate is admitted into AGAP. According to the IAAP's Constitution ..., a member of one of IAAP's Group Members automatically becomes a member of the IAAP.[42]

When in November 2005 the Zürich court resolved the lawsuit in AGAP's favor, it also upheld the AGAP members' approval of the new constitution and the delegation of AGAP's training right as well as the IAAP's position on these matters. While the complete court judgment in German[43] was posted on the website, a summary in English was also made available, excerpts of which are included below:

Overview

In the fall of 2004 AGAP members Ernst Spengler, Brigitte Spillmann and Robert Strubel jointly filed legal claims against AGAP, on the whole attacking the legitimacy of the new Constitution that had been approved by circular vote during the summer of 2004. The obligatory meeting with the Justice of the Peace in November 2004 failed to bring resolution. Consequently, between February 2005 and the end of October 2005, there followed the orderly exchange of legal briefs.

On November 16th 2005 AGAP's President received the Court's Judgment in which it was stated that Plaintiffs' claims were fully rejected. On November 28th, commensurate with the end of the appeal deadline, the AGAP Executive Committee learned of Plaintiffs' decision to forego appeal. The Court's Judgment thereupon became final. In particular the final Judgment upholds AGAP's new Constitution, AGAP's constitutional right to conduct training as is now realized in ISAPZURICH, and the correctness of all Executive Committee procedures leading up to and connected with the votes on these items.

Summary Claims and Court Findings

[....]

9. In the initial Statement of Claim Plaintiffs defended their allegations with lengthy reference to the on-going conflict at the CGJI. In this context they alleged that the real aim of the AGAP Executive Committee in proposing the new Constitution was to collaborate with other unnamed "proponents" to destroy the CGJI. The court dismissed this claim along with all reports on the conflict development with the Curatorium, ruling that these are irrelevant to the Association Law under which the Statement of Claim was presented.

10. Having lost the case Brigitte Spillmann, Ernst Spengler and Robert Strubel are now ordered by the court, in accordance with Swiss law, to jointly pay CHF 11,000 indemnification to AGAP's attorney—and in addition taxes and court costs, together amounting to nearly CHF 20,000.[44]

For Dr. Markus Wirth's stellar defense of AGAP in the lawsuit, we are eternally grateful. The professional guidance and support he gave us was sorely needed after the years of grueling conflict and disappointment. Stacy Wirth, then AGAP Secretary, provided irreplaceable assistance in the process. The legal victory gave ISAP the confidence it needed to sail on its way toward a new and promising future.

Have we found our true north at ISAP? Have those colleagues who remained at the Institute found their true north? It is an ongoing process to be sure. ISAP and the Jung Institute are now very different from each other in "ways of training" and yet, as all trainings worldwide, both institutes in Zürich are based on Jung's original concepts and writings and centered on the principle of individuation. Perhaps the groups needed the split to individuate, each in its own way. Perhaps those involved in Zurich training are still in a process that has not reached its final conclusion.

One thing is for certain: Without the convergence of many streams, flowing with a wide variety of emotions from the most base to the most lofty, ISAPZURICH would never have come to be. The founding was made possible by living through disappointment, intense soul searching, and coming to conviction. ISAPZURICH, born of necessity, was the best that could have emerged in the fall of 2004.

NOTES

[1] The Curatorium, C.G. Jung Institute, Zürich/Küsnacht, in a letter to all analysts, patrons, and employees of the C.G. Jung Institute dated September 22, 1998.

[2] SOS venues included the homes and the offices of Andreas Schweizer, Hanna Hadorn, Diane and Paul Brutsche, Stefan and Sasa Boëthius, Ursula Ulmer, Freya Bleibler, and Helmut and Ellynor Barz.

[3] The Editorial Board of *The Bridge* consisted of Deborah Egger, Constance Steiner-Blake, and Maxine Schmid-Matison. Contributors were Vicente de Moura, Art Funkhouser, John Granrose, Dennis Gyurina, Gary Hayes, and Douglas Whitcher. The layout was overseen by Franziska Lang-Schmid.

[4] AGAP's original Constitution (1954) provided not for a "president," but for an equivalent "executive secretary." In written correspondence the term sometimes converted to "honorary secretary." The term was replaced by "president" in August 2004 and, for clarity's sake, we present that title from here forward.

[5] Hans-Rudolf Kuhn writing in *The Bridge,* No. 2, 1998, p. 9.

[6] Our late colleague, Ian Baker, aptly associated SOS with Save Our Souls.

[7] Brigitte Spillmann in *Info 1-2000 / Curatorium, C.G. Jung Institute Zürich,* December 1999. The relevant text in full reads as follows:

As you know, the voting for the persons to replace the three members who left the Curatorium at the end of last year [1998] was with the explicit provision that the Curatorium would conduct a consultative re-election vote for the entire Curatorium at the end of the 1999/2000 Winter Semester. This obligation, which both the remaining and the newly elected members agreed to, will now be fulfilled by the Curatorium.

See also, "Important Dates: 24 March, Consultative Re-Election Vote for the Curatorium," in *Info 2-2000 C.G. Jung Institute Zürich*, March 2000. 20000300_Calendar_Consultative Election.pdf.

[8] Brigitte Spillmann and Robert Hinshaw speaking for the Curatorium in "To the Analysts, Students, Patrons, Donors, and Staff of the Institute" on April 7, 2000. 20000407_Cur re Consultative Election March 2000.pdf.

[9] Paul Brutsche, "An die Analytikerinnen und Analytiker, die StudierendenvertreterInnen, die Mitglieder des Patronats, die Mitarbeiterinnen des C.G. Jung-Instituts" on April, 12, 2000. 20000412_PB betr Konsultativwahl März 2000.pdf.

[10] Brigitte Spillmann, "Persönliche Erklärung...Mitgliederversammlung der SGAP in Basel, 20 Mai 2000." 20000520_BS an MV-SGAP.pdf, our translation.

[11] Ueli Mühlemann, "Status Analysis for the Attention of the Curatorium, Mandate of mundi consulting ag from the C.G. Jung Institute, §3 Work Results" on February 8, 2000 and appearing in *Info 2-2000 C.G. Jung Institute*, March 2000. 20000300_Mundi Work Results_CGJI-ZH.pdf.

[12] From the Bundesamt für Gesundheit BAG, Direktionsbereich Gesundheitspolitik, "Psychologieberufsgesetz (PsyG), Qualitäatsstandards § 3.2 Selbsterfahrung" on January 1, 2014.

[13] Daniel Baumann, "Commentary on the Annual Financial Report 2000/1, Beginning of December 2001," from *Info 3-2001 / Curatorium C.G. Jung Institute*. 20011200_Cur Commentary on the Financial Report 2000-2001.pdf

[14] Ibid.

[15] Antoinette Baker and Hans-Peter Kuhn from originally unpublished concerns transmitted to the Curatorium and discussed in several private meetings.

[16] Brigitte Spillmann, Daniel Baumann, and the Curatorium, "The C.G. Jung Institute Seeks Your Support!" March 3, 2003. 20030303_CGJI-ZH Seeks Support.pdf.

[17] Ian Baker, Paul Brutsche, John Hill, Lucienne Marguerat, and Maria Meyer-Grass, "An die Mitglieder des Ausbildungsrats," March 10, 2003. 20030310_AURA Vorstand an Mitglieder.pdf, our translation.

[18] In "Vereinbarung zwischen Irene Lüscher, Brigitte Spillmann, Felix Huber, Ernst Spengler, Kläger und Schweizerische Gesellschaft für Analytische Psychologie, Beklagte, betr. Anfechtung eines Vereinsbeschlusses," June 2, 2003. 20030602_Klage_Vereinbarung IL, BS, FH, ES, SGAP.pdf.

[19] Brigitte Spillmann and the Curatorium, "Declaration of Affiliation," July 15, 2003. 20030715_Declaration of Affiliation CGJI-ZH.pdf.

[20] For details, see Brigitte Spillmann and Irene Lüscher speaking for the Curatorium in "To the Analysts Active in Switzerland" on September 12, 2003. In this letter—dated one day in advance of the Erlengut meeting—the Curatorium pronounced, "Until further notice, contributing payments will be considered as an *equivalent* to a Declaration of Affiliation [their italics]." The payment deadline was now extended from September 1 to October 1. Most crucially, however, as "recapitulated" here, the "consequences" for non-payers remained the same.

[21] See Ernst Spengler, "What is Going on at the C.G. Jung Institute?" as seen in Ernst Spengler, Irene Lüscher, and Brigitte Spillmann, "The State of the C.G. Jung Institute Zürich: Conflicts, Development, Institutional Dynamics, A View as Seen from the Curatorium's Perspective," January 2004. Here the Curatorium conceded that,

> Because of the analyst's lack of formal binding it has not been possible for the Curatorium to bind them to a duty of [financial] contribution. A request [for payment] with sanctions following a non-response would have meant coersion by law and therefore was not possible to enact.

[22] Stefan Boëthius, Paul Brutsche, Katharina Casanova, Deborah Egger, Hans-Peter Kuhn, Doris Lier, Lucienne Marguerat, "Brief an die Studierenden des C. G. Jung Instituts-Zürich," September 15, 2003.
20030915_Erlengut_Brief an Studierenden_Sept 2003.pdf.

[23] See, for instance, Brigitte Spillmann, Irene Lüscher, and the Curatorium in "Einschreiben, Deborah Egger" on December 11, 2011. 20031211_Curatorium Verweis_Beispiel.pdf.

[24] Deborah Egger in response to Murray Stein, "Re: Your letter of September 17, 2003, 'The Catalogue of Affiliated Analysts'" 20031218_Egger-AGAP to Stein-IAAP.pdf.

[25] Murray Stein with former IAAP Presidents Adolf Guggenbühl-Craig, Hans Diekmann, Thomas Kirsch, Verena Kast, Luigi Zoja and President-Elect Christian Galliard, "Letter to the President of the Curatorium of the C.G. Jung Institute Zürich, the President of SGAP, the President of AGAP, and to All Parties and Individuals Involved in the Present Crisis," September 17, 2003. 20030917_MS et al to Curatorium President et al.pdf.

[26] ETH Zürich is a "leading international university for technology and the natural sciences. It is well known for its excellent education, ground-breaking fundamental research and for implementing its results directly into practice." Accessed July 6, 2014, http://www.ethz.ch/en/the-eth-zurich.html.

[27] Brigitte Spillmann, Irene Lüscher, and the Curatorium, "Curatorium Decisions Related to Rector Osterwalder's Mediation, Enclosure 1, point 2," January 29, 2004. 20040129_ Curatorium Decisions re Mediation.pdf.

[28] Marco Della Chiesa, "Verhandlungen mit dem Curatorium," April 26, 2004. 20040426_SGAP_Verhandlungen mit dem Curatorium.pdf.

[29] Brigitte Spillmann, Irene Lüscher, and the Curatorium, "The Board of SGAP Breaks off the Negotiations [Der Vorstand der SGAP bricht die Verhandlungen ab]," May 7, 2004. 20040507_Curatorium on Negotiations.pdf.

[30] Konrad Osterwalder, "An die Analytikerinnen und Analytiker des CG Jung-Instituts Zürich," from early May 2004. 20040500_Osterwalder_Ende Verhandlungen.pdf.

[31] Konrad Osterwalder, "Stand der Verhandlungen mit dem Curatorium," as seen in Marco Della Chiesa, "Protokoll der ordentlichen Mitgliederversammlung der SGAP vom Samstag, 15. Mai 2004," §7, May 15, 2004. 20040515_SGAP_Protokoll_MV Mai_Stand der Verhandlungen.pdf, our translation.

[32] The following colleagues were affiliated with the SOS group: Peter Ammann, Kathrin Asper, Antoinette Baker, Ian Baker, Nathalie Baratoff, Ellynor Barz, Helmut Barz, Freya Bleibler, Sasa Boëthius, Stefan Boëthius, Vreni Osterwalder-Bollag, Katharina Casanova, Diane Cousineau Brutsche, Paul Brutsche, Maria Danioth, Rossana Dedola Roth, Brigitte Egger, Deborah Egger, Dirk Evers, Karen Evers, Hanna Hadorn, Maria-Elisabeth Hartung, John Hill, Franz-Xaver Jans-Scheidegger, Ursula Jaquemar, Ute Jarmer, Waltraut Körner, Hans-Peter Kuhn, Doris Lier, Kari Lothe, Sonja Marjasch, Urs Mehlin, Alice Merz, Cedrus Monte, Annemarie Moser, Stéphanie Odermatt-Edelmann, Elisabeth Peppler, Bernhard Sartorius, Jody Schlatter, Sandy Schnekenburger, Andreas Schweizer, Constance Steiner, Ursula Ulmer, Ethel Vogelsang, Joanne Wieland-Burston, and Stacy Wirth.

[33] From the "AGAP Executive Committee Statement on the Zürich Institute Situation, January 2004," *AGAP Newsletter,* March 2004, p. 8. 20040300_AGAPNL2004.pdf.

[34] The Curatorium, "Change in AGAP Bylaws: Comments from the Jung Institute of Zürich," June 2004. 20040600_Curatorium on AGAP Bylaws.pdf. For AGAP's response see, for example, Deborah Egger, "RE: Comments from the C.G. Jung Institute on Changes in AGAP Bylaws" in August 2004. 20040800_DE Answer to the Curatorium Letter of June 2004.b.pdf.

[35] Murray Stein, in a letter to Brigitte Spillmann, July 26, 2004. 20040726_MS to Spillmann July 27.pdf.

[36] Meeting Minutes compiled by Deborah Egger and Stacy Wirth, AGAP Business Meeting, Barcelona, Summary Minutes of Resolutions, §1, from August 30, 2004. See the Appendix, Source No. 0, 20040830_AGAP Minutes BusMtg Barcelona.pdf.

[37] The new AGAP Constitution permitted the inclusion of proxy votes at the General Assembly, which required new and rigorous registration procedures. Without the diligent help of Andrew Fellows and Mae Stolte in Barcelona, the registration of over 100 members and their proxies could not have been accomplished.

[38] Stein, July 26, 2004, §10.

[39] Meeting Minutes compiled by Deborah Egger and Stacy Wirth, AGAP Meeting for the Constitution of the Delegated Training Program in Analytical Psychology, Summary Minutes, §1, from September 9, 2004. See p. 195, Appendix Source No. 3, 20040909_Minutes_AGAP Constitution of Delegated Training_sig.pdf.

[40] See, for example, Brigitte Spillmann, Irene Lüscher, and the Curatorium in a letter to Mr. John Betts, March 3, 2005. 20050303_Curatorium to ISAP Students_ example.pdf.

[41] Ibid.

[42] See p. 204, Appendix Source No. 5, Christian Galliard in a letter to Mr. John Betts, September 13, 2005. 20050300_Galliard IAAP to ISAP Student.pdf; see.

[43] From Bezirksgericht Zürich, Prozess Nr. CG050030/U, 3. Abteilung, "Urteil vom 11. November 2005 in Sachen Ernst Spengler, Brigitte Nelly Spillmann-Jenny, Robert Joseph Strubel, Kläger gegen AGAP Association of Graduate Analytical Psychologists ...," November 2005. 20051111_Bezirksgericht ZH_Urteil_Nov 20050.pdf.

[44] Deborah Egger speaking for the AGAP Executive Committee in "Summary of the Judgment in the Lawsuit Against AGAP" in November 2005. 200511_Summary of Judgment November 2005.pdf.

(Most sources are available in the ISAPZURICH library.)

OPENING SPEECH ISAPZURICH
OCTOBER 23, 2004

PAUL BRUTSCHE, ISAP PRESIDENT

Dear Colleagues,
Dear Students,
Ladies and Gentlemen,

I.

It is a great pleasure for me to welcome you to this opening party on behalf of the Officers Committee of ISAP. It is wonderful that so many of you could come. It gives us the opportunity to test the capacity of our new home... and also to test the nerves of the people working for the fire brigade. But above all, it shows how much interest and support there is for ISAP.

The fact that we are able to begin with the very first semester of our new training program today is nothing less than a miracle. After an accelerated pregnancy period of 5 months, the tangible results of this complex organization have been born. It was only possible with a lot of selfless dedication on the part of number of individuals and with the support of the cosmos in the form of favorable circumstances.

Let me mention some of the favorable circumstances and some of the miracle workers.

First of all, I would like to mention our landlord, Mr. Maissen (who unfortunately could not be with us tonight). Mr. Maissen is the owner of this building. As soon as he heard about our project, he was so enthusiastic about it that he spontaneously offered to rent us these two floors under very favorable conditions. He still has his personal office on the top floor so that his good energies will continue to beam down on us from above.

Many sponsors have also contributed to make our project possible. We have received spontaneous donations from colleagues and friends, ranging from 1,000 to 100,000 francs. Such generosity has permitted us to realize a vision that would otherwise have had to remain a beautiful, but unobtainable, dream. We are extremely grateful to all these donors.

The first colleague to whom we owe our very special gratitude is, of course, Debbie Egger, the President of AGAP. Debbie was the initiator of the project and since then has acted like an engine pulling it along through the phases of its development, thanks to her incredible capacity for hard work and clarity of mind but also thanks to her deep faith in the value of our Jungian heritage. The fact that we can gather together here tonight is mainly due to her. If your hands happen to be free, I would suggest that we give Debbie a warm round of applause.

We are also indebted to our lawyers: Martin Amsler, who has been involved in the project since the very beginning, developed its legal structure. As the lawyer of both AGAP and IAAP, his years of experience have permitted him to help us navigate successfully through the sometimes turbulent waters of all the legal aspects surrounding such a project. His enthusiasm for and dynamic commitment to our project has been an invaluable asset. We are also deeply indebted to Dr. Markus Wirth, who has committed himself to defend our newborn child against all those who intend to threaten it, already, in the early moments of its life. Out of love for our initiative, Dr. Wirth agreed to defend our cause, charging fees that would be considered starvation earnings among lawyers.

Now every individual who is seeking a role in the collective needs a good persona. Because of this, ISAP had to present itself with an attractive logo as well and a set of rules and regulations that would not only *be* serious, but *look* serious. This requires professional help, which is usually very costly. We were lucky enough to find someone, in the person of Marcus Baker, who offered to do this for us "pro bono." Not only is Markus very competent in his profession but, as the son of Antoinette and Ian Baker, he already knew in advance what to expect from a working relationship with Jungian analysts, and how nerve wracking such an endeavor is.

And now, of course, I have to mention those who have been directly involved in the shaping of ISAP in its multiple facets. These are the members of the Officers Committee. In order not to make this into an endless and boring speech, please allow me to say just a few key words about each of them. At The same time this gives me the opportunity to introduce them to you, in case you don't already know them. Stacy Wirth is Vice President; she writes like crazy detailed minutes of our weekly meetings and is responsible for legal matters of which we unfortunately have no shortage. Stefan Boëthius is our treasurer who, with a steely determination, doesn't allow us the least unnecessary expense. He is very knowledgeable in Internet matters and deals with our internal and external communication. He is also the person who has created Info-Forum, which has proven to be a wonderful platform for a free exchange of views between colleagues. Karen Evers is Director of Administration, who altruistically offered herself for this extremely demanding job and who manages to fulfil it with admirable and nearly inexplicable calmness. Katharina Casanova is our Director of Studies and has written more emails in these last few weeks than in her whole life. She deals daily with very complex questions and, in the midst of all this, takes the time to create a good atmosphere of personal and mutual trust between ISAP's Committee and the students. Nathalie Baratoff. If Hercules had been a woman, her name would probably sound a bit like "Nathalie." Alone, she managed to set up the whole program of lectures and seminars for the winter semester. She is also investing her energy in creating a library for ISAP. Doris Lier is Director of the Selection Committee. But she is also the person we have to thank for getting us the offer of this ideal location. For several years she has worked in this building and obviously left such a positive impression that its owner was convinced that he could rent it to us without fearing that he would be invaded by a horde of freaks.

Now that I have talked about the official delegation, I would like to mention some less visible but, nonetheless, essential contributors. Out of a sense of justice, I will name them in alphabetical order: Kathrin Asper has helped us immensely on a number of occasions with her input. She doesn't want her name to be mentioned and that is why I am not mentioning her.

Tonie Baker. As a former president of the Selection Committee, she played an important role in initiating our new Selection Committee Director into the many secrets of the function. She has also, together with Hans-Peter Kuhn, taken charge of compiling our new regulations. Together they have accomplished a huge and extremely complex task: having to create a set of regulations that would be compatible with those of the old Institute, while being other, a paradox they have managed to master admirably. Sasa Boëthius. In the last few months she has hosted several of our preparatory meetings, and spontaneously offered to organize this opening party. So the warm and friendly atmosphere that we will be enjoying tonight is largely due to her. She has been assisted in this task by Monique Wulcan. Given the limited space available in our new ISAP home and the constantly increasing list of our guests, the whole thing must have felt like an irresolvable logistic problem. We are very grateful that they didn't collapse under a panic attack. Diane Cousineau Brutsche. This person I know relatively well and I have been in a good position to witness not only her active co-operation as Vice President of AGAP since the very beginning of the project but also her intellectual, emotional, and culinary support of our team through the development of the work.

Urs Mehlin is someone to whom we always come back when there is a public meeting to lead. Urs has also constantly shared his invaluable reflections and insights with us. And last but not least, he has been extremely helpful in his ability to find us cheap and good technical equipment. And on top of that to make it even cheaper for us, he came last week in a car fully loaded with equipment he is donating to us. John Hill is another moderator whom we also always appreciate a lot. Together at the constitutional meeting, he and Urs managed to bring forth a cosmos out of what was promising to develop into a huge chaos. And John, as a long-standing Executive Committee member of AGAP, has also been a very precious collaborator in the development of our project.

And now of course, I am afraid that I may have forgotten several colleagues whose names I am sure will come back to haunt me in the middle of the night and give me a few sleepless nights.

That reminds me that I have indeed forgotten to include three colleagues who have played an essential role in the creation of ISAP and without whose assistance ISAP would certainly never have come into existence. Even if they are not expecting any special mention, I feel the need to mention our colleagues from the Curatorium: Brigitte Spillmann, Ernst Spengler, and Irene Lüscher.

II.

Now, if you permit me, I'd like to say a few words about some of the labor pains that have accompanied ISAP's actual birth process.

It is certainly not difficult to imagine the gigantic efforts required in getting this project up and going during the last few months. The progress of our work could be monitored by a quick look at each other's faces, which got increasingly grey and pale with every day that went by.

And then, from time to time, we were haunted by a dreadful fantasy. As the number of colleagues ready to offer lectures and seminars was rapidly increasing without us having the slightest idea how many students (if any) would enroll in our program, we were confronted with the possibility of seeing all our hard work and sleepless nights leading to a Kafkaesque nightmarish scenario with 60 enthusiastic analysts swooping down upon 2 or 3 helpless students who would be buried alive under a wealth of theory and good will. Thank God, it turned out differently.

Among the innumerable challenges we had to face, let me pick out one as an example: How should our new training program be named? A difficult task that any parent confronts when a newborn baby enters this world. One searches for a name that doesn't sound too exotic but, at the same time, not too banal. It is also important to avoid choosing a name that will remind one of an old, shrewish aunt or of a former lover. The problem was the same for us.

In concrete terms, there was no way the newborn baby could be called C.G. Jung Institute, even if this was the name that we would have readily chosen. For legal reasons, the name Jung was also not allowed to appear, as it would have been likely to generate confusion. So we had to find a name for a Jungian training program that wouldn't make it sound like a Jungian training program. The only solution was to use the term "analytical psychology." So we opted for this name while letting the term, Jungian, creep into the subtitle, in script small enough to hopefully escape the eyes of a short-sighted lawyer.

For similar reasons there was no question of using the word "Institute." So how about "Seminar?" In German, it sounds wonderful, but in English, much less serious. What about "School?" In English it is acceptable, but not in German. Even in English, "School" had to be polished with the adjective "international" to bring it closer to the major leagues. Then we had the problem of inserting the word "Zürich" somewhere. But where the hell to put it? Zürich School of Analytical Psychology? But then the word "international" would have got lost, defeated by "Zürich." This was unthinkable: both words had to be included but had to find their rightful place. So, "International" was chosen as a bright prelude at the beginning, while Zürich was relegated to the end of the name, as a sort of modest rearguard.

By this time we had acquired two names, a German one and an English one. But both of them were so long that at the beginning we had to constantly refer to our written documents to remember what our actual name was. Then came the moment of compression. ISAP was the solution, especially as it fitted the name in both languages. But what about "Zürich" at the end? IS-APZ? No human tongue would have ever been able to pronounce this decently, even those who are practiced in using the Zürich dialect. So "Zürich" at the end had to be dropped off from the compressed form.

Anyway, to close this chapter, I beseech you; I implore you, for heaven's sake, don't make any further name proposals. None of us would survive it.

III.

But let me come now to matters concerning the present and the future of ISAP. First of all, I want to turn my attention to our students. I feel the need to not only welcome you wholeheartedly to ISAP but also to congratulate you for having been able to come to a decision that was in no way a foregone conclusion. It has demanded courage and faithfulness to your own inner process to branch off the familiar path you had been following until now, to renounce the outer status you enjoyed as students at an Institute renowned the world over and to enter into a new experience with all the uncertainties that belong to any new path. We feel enormous respect for your authenticity, your inner flexibility, your readiness to take a risk, and your trust. We are also aware that it has demanded that you go through a painful grieving process to leave behind an Institute in which you had invested so much soul energy and loving fantasies.

We thank you for the trust you put in us.

We fear your expectations.

We count on your patience.

And we encourage you to participate actively in the shaping of our new training program.

Rest assured that you will be heard and that your suggestions and initiative will be most welcome. Together we will be able to create a warm and pleasant environment and a feeling of community. Together we will bring new ideas and inspiration to Jungian training.

IV.

And now, before ending, here are a few points concerning a possible vision of our school for the future.

The foundation of a new training institution in Zürich should not be a simple cloning of the one we have known and cherished in the past. In that case, this long and painful crisis would have served no purpose. So we want to grasp this opportunity to develop a new movement in our training, be it in its shape or in its content. There should be room for new experiments and innovative ideas. We do not intend to be a mere reproduction of the Institute, a sort of bonsai version, a Seehof—without the lake.

In order to succeed, we of course need the active involvement of all those who have chosen to participate in our new program. So we have to take advantage of the democratic structure that the AGAP Zürich sub-group offers us. We want to develop a form of leadership that will be supported and influenced by all of those who are actively involved in the training. Our democratic structure is the best tool for the development of a vital sense of community.

And finally, we intend to make this a place where our Jungian heritage will be respected. Our Zürich tradition is unique in the Jungian world, and we want to maintain this unique heritage. But we also want to keep it alive and this means being open to the contemporary scientific and spiritual developments. This new home of ours, despite its rather modest dimensions, or maybe even because of that, could very well be the ideal vessel for such new growth to take place.

In this spirit, I wish you, and each one of us, a good start in our new school: To you, our students, I wish a fruitful process in an inspiring atmosphere. To us, analysts, teachers, helpers and sponsors, I wish the kind of satisfaction resulting from the awareness of being involved in an exciting, creative joint enterprise in which each person can make a precious and valuable contribution.

REMARKS ON ISAP'S OPENING
OCTOBER 23, 2004, HOCHSTRASSE 38, ZURICH

DEBORAH EGGER, AGAP PRESIDENT

On behalf of AGAP, I welcome you to the opening of the new International School of Analytical Psychology. AGAP, an organization with a membership of around 650 Jungian analysts all trained in Zurich and now living in 27 countries around the globe, is the structural "parent" of this training program. AGAP is the largest Society of the IAAP, the second largest in voting strength, and its size means that one out of every four Jungian analysts in the world is an AGAP member. AGAP is a charter member of the IAAP, the international umbrella organization, I will say a little more about AGAP in a few minutes, but first I want to put our being here in context.

As I thought about what we are doing here tonight, I realized with great clarity that we are not here, really, out of complete free will. We're not here out of spontaneity and not by accident or by happenstance. We are here because it is necessary to be here. That means that this school's existence is inevitable, its creation has been compulsory and was absolutely needed.

To be born out of necessity means to be born under pressure of circumstance or from physical or moral compulsion, out of a state or quality of need and impoverishment. To be born out of necessity also means to be accompanied by urgent desire, in such a way as to feel that it cannot be otherwise.

And here we find ourselves tonight, inaugurating a new training institute, born of necessity, under pressure of circumstance and accompanied by urgent desire, but with an absolutely unmistakable enthusiasm. An enthusiasm I experienced first *en masse* on July 6th when, for the first time, a viable plan for a new training program was offered to analysts. An enthusiasm I felt since then from students as they heard there could possibly be another legitimate option for their studies in Zurich in an atmosphere of warmth and respect. And an enthusiasm that brought tears to my eyes in Barcelona as colleagues from around the world (and not only AGAP colleagues) shared their support and interest with remarkable energy. When I was reminded that the root of enthusiasm was the Greek word, *enthousiasmos,* and it originally meant to be inspired, I understood, in a new way, the powerful wave of energy that has been moving this endeavor along.

In addition to being born of necessity, this new school is enthused with inspiration. It is inspired in a most inexplicable and mysterious way. We now have the daunting task, each of us, each student, each analyst, each teacher, each supervisor, each volunteer in whatever capacity, to embody this inspiration, this spirit, of the work on the human psyche that can only be done this particular way in this particular place.

It is indeed an historical moment that conditions at the Zurich Institute, which Jung himself founded and which has been such a magnet for each of us, would necessitate (force, compel, require) this many people to come together and begin a new school of analytical psychology in Zurich. But obviously the need for renewal was ripe and we are the torchbearers of that renewal.

My personal words of dedication for us tonight come from the poem that showed me I had found my heart's home in Zurich, at a moment 18 years ago in the C.G. Jung Institute. I have recently learned that this poem, which meant so much to me then, was written by one of the most irreverent and iconoclastic Zen masters of medieval Japan. In an interesting way, this revelation makes it even more fitting for this dedication than I had first realized.

The Silver Thread
Following rules inflexibly
Produces fools.
Contradiction is the essence of
Human consciousness.
Rules, rules innumerable

As grains of sand on the shore
Confuse the spirit.
From birth onwards
The only certain guide
Is the silver thread.
Every spring
Blossoms open,
Only to close again.

— Ikku (Zen poem), also known as "Crazy Cloud" (1394-1481)

And what a paradox and contradiction (which as the poem reminds us is "the essence of human consciousness") that this poem which once adorned the official documents of the Jung Institute still holds a valuable message for us as we take leave of that very place, carrying with us the spirit, the essence, of a training institute which rests in the larger framework of the "work on the human psyche," and which ultimately (as we all know so well) cannot be cemented by rules.

And it is with utmost humility and absolute faith, I hope, that we continue to follow the "only certain guide," our own individual silver threads, always remembering "every spring blossoms open, only to close again." And this out of necessity and enthused with inspiration.

In closing, I'd like to shift the focus back to AGAP. The fact that AGAP is the "carrier" of this training will have a great impact on the project, I believe. It gives an opportunity for many of us (who could not stay at the Jung Institute) to continue training in Zurich in a spirit that has been referred to as the "Zurich Tradition" and recognized by the IAAP. But there is another important opportunity here as well. And that is to bring new vibrancy and enriched meaning into that "Zurich tradition" through a lively and creative exchange with our colleagues around the world who care deeply about the privilege of training in Zurich.

We stand on the verge of a new relationship with our colleagues and supporters in the international community. We have Jungian groups interested in engaging with us, proposing a variety of ways of cross-fertilization in South America, in North America, in Asia, and in other parts of Europe. These colleagues look to Zurich as the "Mother/Father Source" (as Mecca) and I believe we can only confirm this image of strength and resourcefulness (Quelle) by being open to new ideas, by exploring possibilities, and by taking risks to venture into unknown territory in Jungian studies and training. (This of course includes any creative cooperation with the Zurich Institute, which may come into being.) Our learning curve is steep and we won't get it right all the time but, with tenacity and integrity, with trust and patience, I know we can meet the challenges we face with clarity of spirit and peace of mind not to mention with heart and soul.

Now before I give up the floor, I must say a few words about Paul Brutsche. We are extremely fortunate to have him as ISAP's first president. Before the elections of officers on September 9th, Paul and I spoke very bluntly about the delicacy of a former president of the Curatorium running for President of ISAP. This was also an open topic of debate at the elections themselves, but in the end, Paul was elected simply because, as one colleague at the meeting put it, "We need the best person for the job to get us off to the right start." He brings a wealth of experience and knowledge to this endeavor, which we would only really understand if he were not here. But besides that, he brings himself fully and without reservation with openness, fairness and self-reflection. (And, as you already know, with a steady sense of humor.) Would you help me thank him with a warm round of applause?

OUR IDENTITY

WHY ISAPZURICH IS UNIQUE IN ALL THE WORLD

MURRAY STEIN

It is bold and perhaps foolhardy to claim ISAPZURICH, is unique but I will not let that deter me. This is because I believe it is true. There is no other Jungian psychoanalytic training program like it in the world today. I will explain why but, for now, trust me, it's true.

I will not allow myself, however, to claim that the training offered at ISAP is the best in the world, only that it is unique. Who can say what is "best"? But this is sure: For some people, ISAP is the only Jungian training program around that effectively allows its students the opportunity to enter into a deep process of formation as an analyst. This formation takes place over the course of several years fully immersed in the analytic culture located now at the old post office building at Stampfenbachstrasse 115 in Zurich.

I.

What makes ISAP unique? I will come to the point: It's in the *prima materia* that is found at ISAP. That is to say, it's all in the ingredients and in the totality of all the ingredients. A finished meal depends on the ingredients that go into it, and its quality is finally a product of these and the artistry of the cook. The distinctive feature of ISAP is in the ingredients first of all, secondly in the distinctiveness of its programs, its curriculum, and its model of total immersion, and thirdly its *vas*, the magical container of Zurich, the city where the process is located and grounded.

First, then, are the ingredients, the *prima materia*. From Jung's writings on alchemy we know that the final product sought—the *lapis*, the "philosopher's gold"—lies hidden and contained within the materials that are assembled and placed into the *vas* at the beginning of the *opus*. Often these materials do not look very promising at the outset, but it is essential to get the right combination of metals and minerals, plants and liquids, and there are many recipes for this. At ISAP, the prima materia is made up of the people who learn and participate in the programs—the students and their teachers. Among these, alchemy is at work from the beginning to the end of training and beyond. Students interact with one another, they interact with their teachers and personal analysts and supervisors, and they interact with the unconscious that emerges at many places in the course of their studies. This combination is unique at ISAP because the students and teachers and training analysts come from so many different countries (Some 20 lands are represented in the student body at any given time; the analysts, likewise, hail from many countries and cultures.) and from a wide variety of backgrounds (the sciences, the arts, business, and other professions). This tremendous variety of cultural and educational backgrounds and ages (from people from 20 to 80) creates a very rich and, indeed, a unique, prima materia at ISAP. ISAP is multiculturalism realized; it is globalism in action. It is the world in miniature and interactively engaged across all boundaries and borders.

Take a glimpse into ISAP. I open a seminar I am going to lead on "A Comparison of Freud and Jung on the Psychology of Religion" by asking the twenty or so students to identify themselves briefly by background and country. In a variety of accents, some clear and others somewhat opaque, they answer in English: They hail from Canada, Russia, Israel, USA, Switzerland, Finland, Taiwan, Japan, Brazil, UK, Ireland, Chile, and Germany. Canadians are the most numerous with three. No country or culture or language group is dominant. The religious backgrounds are spread across a wide spectrum. The professional backgrounds are also diverse–psychiatry, business, clinical psychology, social work, religious studies and divinity, medicine, art therapy, political science, economics, and philosophy. This could be a Tower of Babel with so many languages and perspectives speaking at once and each unintelligible to the other, but it is not. The perspectives enrich, the cultural differences contribute their flavors, and archetypal images and themes emerge from the burbling conversations. There are moments of insight and moments of emotion. There are even numinous moments for we are, after all, dealing with religion and religious experience. Jung's massive breadth of interests and references can encompass all the cultures brought together here. Each student gives and takes what she or he can, and they take it back for further incubation and reflection. As seminar leader, I am moved by the material and by the psyche that collects and gathers in the room over the four-week period of the seminar. I believe no one leaves this seminar untouched. Each person contributes, either by speaking or simply with intensive listening and presence.

II.

Second, the training is seamless at ISAP between classrooms and analytic consulting rooms. What happens in the seminar enters into the analysis; what happens in analysis comes back subtly into the classroom; what emerges in dreams and active imagination passes through the analytic process into the classroom as depth of perspective and psychological insight; what leaps into the mind from the reading of psychological texts feeds into the classroom and into the analytic process; what sparks insights in class is further processed in supervision of control cases. And so it goes within the *vas* of ISAP training. And this grows over a period of years. It is intensive and formative. That is training. The incredibly rich program—made up of sixty offerings per semester that include didactic and experiential, clinical and theoretical, onsite and offsite learning experiences and locations—fills the vessel of alchemy to the brim. Fires heat the material, psyche cooks, and the peacock shows its feathers.

The training and supervising analysts at ISAP also powerfully enter the process with their contributions. They do not stand apart and at a distance. They are also in the vessel of transformation and on site. They bring their own psychic materials into the formative process of training. Their contributions take place in three principal locations: in personal analysis, in the supervision of cases, and in the classroom. The uniqueness of ISAP arises from the personal analysts, supervising analysts, and teachers who are all on site and in continual contact with the students. They are in the *vas*, not somewhere standing at a distance and contributing their bit in absentia. The full immersion model of training at ISAP includes the teachers and training analysts. They are immersed as well as the students. They, too, are part of the prima materia even as they oversee the process as it develops for the individual candidates. In other words, ISAP is an intense learning community on a full-time schedule.

It is different from an academic learning community, however, for it is concerned with more than cognitive mastery of materials. Its main objective is to form good Jungian psychoanalysts, which involves a deep engagement with psychic processes at conscious and unconscious levels. Individuation is the model of this formation. This includes engaging cognitive capacities and the mastery of intellectual materials to be sure, which are carefully tested for in a serious round of examinations at the midpoint of training and at the conclusion. But the formation process also, and more importantly even, pertains to the individual candidate's journey toward wholeness. This cannot be measured in the classroom only or in written work, such as the symbol papers and the final thesis. It is measured on an alchemical scale: Is the transformation process well advanced by the time a candidate graduates from formal training? Has the candidate experienced the unconscious in depth by wrestling with core complexes and engaging the deeper layers of the psyche in archetypal images and impressions? ISAP is keen to train the whole person and not only to instill analytic techniques.

III.

The third essential factor in the uniqueness of ISAP is its location in the heart of Zurich. It is the only Jungian psychoanalytic training program in the city where Jung himself lectured, taught, and analyzed patients. Zurich is the *vas* in which ISAP as a whole is contained. Why does this make a difference? For the answer to this question to make sense, you have to experience Zurich over an extended period of time, not for just a short visit. There is magic here. Partly, this is due to Jung's continuing presence as a *spiritus loci*, which, for Jungians, is a highly important factor, but it is also due to the architecture and age of the buildings, the river that runs through the old city, the church towers and their precious bells, the guild houses, the crooked little streets, the bakeries offering a mind-bending assortment of breads and pastries, the train station, and the city by day and by night in its steady hum of activity. It is an ideal place for introversion during the foggy days of autumn and winter. It is glorious in the spring with flowering trees in the parks and swans on the *Zurisee* and in the *Limmat*. No need for a car. You can take the tram or walk to your analyst's office, to class, or to the library. The city is contained, and it is a container. Many souls have been released to a new life here from their hidebound network of mental habits and complex-burdened egos. ISAP is dedicated to the alchemy of the psyche's transformation in the soul's quest for inner space to grow and manifest.

The students, the teachers, the immersion model, the curriculum, the city – this combination makes ISAP what it is, a training program unique in all the world.

The proof is in the pudding. Does ISAP show results? Do the candidates who advance through the training program and graduate become outstanding representatives of what Jungian psychology stands for and offers to the world? This is difficult to answer. Many return to their home countries upon graduation. Some find a way to stay in Zurich and become new analyst-participants in ISAP. From what I have observed, I am confident that everyone who has passed through this training program successfully and graduated over the last ten years has experienced significant transformation in the course of training. And what I know about the analysts who make up the membership of ISAP and constitute its faculty is that they too are, without exception, engaged in constantl individuation in their personal lives. There is an exciting air of creativity among the members of ISAP, students, and teachers alike. Everyone seems to be deeply engaged in learning, growing, and expressing their gifts and talents in one fashion or another. There is a buzz about ISAP. May it continue for more decades to come!

WHY DEMOCRACY?

JUDITH HARRIS

*The ignorance of one voter
in a democracy impairs
the security of all.*

— President John F. Kennedy

At its best, democracy feeds the soul and furthers the individuation of all that participate. At its worst, democracy breeds inefficiency, chaos, and a spiraling tendency to collapse into entropy. We at ISAP have made a shared commitment to conduct ourselves in as democratic a fashion as possible. What does this mean in theory and what does this mean in practice?

The image we have of democracy in ancient Athens is of free men gathering together in a public square to discuss matters of the moment; rather like the image we have of the citizens of Appenzell gathering together in the public square to discuss whether to include women as equal citizens. Like the Athenians, the Appenzellers continually chose not to include women as voting citizens and had to be forced to do so by the Swiss Federal Government as recently as 1991. Democracy, therefore, has not always meant rule of the people, for the people, by the people because not all the people have had the right to exercise this basic communal activity. Immense political struggles, civil wars, and revolutions ensued when those excluded from the right to vote and participate became sufficiently aware of both their exclusion and their desire to be included and this has been true in all so-called liberal democracies.

In political communities as distant from each other as ancient Greece and the republic of the United States of America, it was also not considered necessary to include slaves in the definition of citizen. The second President of the new American republic, Thomas Jefferson, as liberal and conscious a person as it was reasonable to expect in those days of enlightenment, when reason was triumphant, was himself a slave owner. We now think we know better and we are prepared to enforce regulations that are intended to create a society in which every man and woman of mature and sound mind has the right to express him or herself freely on any issue concerning the community.

In other words, and at least in theory, rule of the people, for the people, by the people presupposes certain conditions: no one shall be barred from participating and voting by reason of race, religion, sex, or economic standing; everyone participating as a citizen has the right to expect as much assistance as is necessary by way of education or access to as much informed, unbiased sources of information as will enable him or her to make a consciously informed choice from the options available, be these at public meetings large and small or at times of decision when vote-taking is deemed to be desirable. It is implicitly understood from the foregoing, therefore, that it is not ethical to try to use any forms of persuasion or coercion to induce a citizen to change his or her mind when it comes to a determining vote. Any forms of bribery, offers of benefits such as a better job, or access to soft options like vacations in the Caribbean should be out of the question.

In theory, then, democracy is obviously a form of government or a way of conducting ourselves within and without, a way which brings forth the most conscious, the most reasonable, the most informed, and the most compassionate awareness of others as our equals.

Democracy requires of us all (and offers to us all) at ISAPZURICH that we see and are seen, that our governance be transparent and open without intermediaries, and that we take individual and communal responsibility—literally, that we respond—for every decision made by us and in our name at the general assemblies. It is incumbent upon us, and it is a privilege, to attend our assemblies; that is, attend in that other sense of waiting. In 1980 a historic and highly contested referendum took place in the province of Québec in Canada. "Yes" votes could have resulted in Québec's separation from the rest of Canada. I was present at a rally in Québec City after the voters had said "No" to the separatist initiative. Many banners proclaimed "J'étais présent au Non"—"I was present at the No." In a living democracy such as ours at ISAP, it is not just an elite few that decide everything for the rest, but everyone can in fact say, "Yes, I was there. We made history that day."

I believe, therefore, that we must all continually evolve in our own process of Consciousness, Unconscious, and Individuation, as Jung himself titled an article in *The Archetypes and the Collective Unconscious* in 1939,[1] to become as informed as possible about each and every item that comes before us as members of the ISAP community and to enter into a dialogue with our own unconscious, while keeping an eye on the collective surrounding us. This applies whether we are analysts or students, it does not matter. As John F. Kennedy said, "The ignorance of one voter in a democracy impairs the security of all." In other words, unconsciousness within our assembly impairs the assembly's ability to serve as a container for us all.

NOTE

[1] C.G. Jung, "Consciousness, Unconscious, and Individuation (1939)," in *The Archetypes and the Collective Unconscious: The Collected Works of C.G. Jung*, vol. 9, §1, ed. Sir Herbert Read, Michael Fordham, Gerhard Adler, William McGuire, trans. R.F.C. Hull (Princeton, NJ: Princeton University Press, 1963).

VOICES FROM ISAPZURICH'S STUDENTS

EDITED BY URSULA ULMER

I am from Canada and Slovakia and I am a teacher and a writer. I am studying at ISAP because the training analyst I wanted to work with is affiliated with ISAP. What impresses me most is the richness of the program and how much hard work and heart the analysts put in.

Eleonora Babejova

I'm from Belgium. Background: philosophy and interior design. Jungian psychology brings together my three interests: philosophy, art, and humanity. What I like most is the richness of multicultural encounters and viewpoints about life.

Nathalie Boëthius de Béthune

I am a psychiatrist from South Korea. I wanted to study Jungian psychology in its hometown. What I like very much is the kind, enthusiastic, loving, welcoming people, lecturers and students.

Jahyeon Cho

I am from Canada, and my background is in music as a singer and singing teacher. ISAP offers the only full-time training program in Jungian Psychology and accepts students from backgrounds other than the medical fields. It offers training that, while looking forward, remains close to the roots of classical Jungian thought. I am impressed by the depth and variety of backgrounds in the analysts and lecturers and the high quality of lectures and seminars. I am stimulated by the community of students, coming from diverse international backgrounds and experience.

Judith Dowling

I am interested in transformation. While living in Canada, I wanted to continue my wholesome relationships with people like the analysts I met at ISAP. What is especially valuable for me is my interpersonal relationships with analysts and our shared focus on the unconscious and individuation.

Sloane Dugan

I came from Japan and I am a psychologist. I study at ISAP because I am interested in the healing effect of images. What makes the training special is the flexibility and originality of lecturers.

Masayoshi Hironaka

I am from Osaka, Japan and I am a former psychiatrist. I wanted to study Jungian psychology, apart from Japan. All lecturers and students are friendly. Lectures are very exciting, something I have not experienced in Japan.

人見佳枝 *Yoshie Hitomi*

I am Canadian, a clinician and health researcher. For a year, I was in a once a month fly-in program and found it insufficient to prepare me for practice in Jungian psychoanalysis and I am already a clinician. What I wanted was a full-time program at the source. What convinced me to study at ISAP is the quality and quantity of lectures/seminars/colloquia, wide choice of analysts, chance to dialogue with senior analysts and international students.

Phyllis Jensen

Above all, ISAP is an experience. An experience of immersing oneself into the realms of an unknown territory, that whether desired or not, will change one's life. That is why I came to Zurich, and to ISAP, for that journey. ISAP

holds the traditional Zurich training that has existed for many decades. Its analysts are astonishingly dedicated to the broad studies of analytical psychology, and above all, to the contemplation, respect, and healing of the human soul. I feel profoundly privileged to be part of this community.
Luis Moris

I am Swiss, but originally from Germany. I worked as an economist for 24 years. I am at ISAP because Jungian psychology has a place for religion. What impresses me most is the training analysis.
Hilde Phan-huy-Klein

For me, ISAP represented an opportunity for the deepest possible immersion in both the theoretical and experiential aspects of analytical psychology. Living in Zurich has itself constituted an important dimension of this experiential immersion, including roots in European heritage.
Stephen Setterberg

I am from New England in the U.S. and have mainly been working as a teacher of literature and writing, prior to and during my ISAP studies. I began study via the Summer Intensive program in Küsnacht where I learned about ISAP. Several individuals especially important to my growth were part of ISAP, and I wanted to follow these threads. Many aspects at ISAP are excellent. Especially inspiring for me is the commitment of the analysts who have formed and guided ISAP's development, along with the quality of the teaching. Helping me to assimilate and integrate this wealth of experience is a pervasive attitude of respect for analytical psychology, its history, and its future.
Margaret Stienstra

I come from Athens, Greece. I have studied history and international relations and have a degree in Law and Diplomacy and worked in Greece, Belgium, Albania, Kosovo and Slovenia in the domains of politics. ISAP offers the unique opportunity of a full-time training in a community of students and analysts from all over the world. What makes the training so valuable is the deep sense of purpose and the commitment to the pursuit of meaning and healing shared by trainees and analysts alike.

Evangelos Tsempelis

ISAP's analytic training has both heart and grit with a variety of aspects for which I am grateful: the absence of rigid curriculum, yet the presence of stringent requirements demanding discipline and integrity; the invitation and encouragement to cultivate a personal ground within a vast field; the curious way the program configures to the shape and trajectory of each candidate; colleagues from a broad range of cultural and academic backgrounds; and the insistent back beat of personal analysis.

As an international student, I am spared the distractions of my life at home and I enjoy the support and finesse of Swiss culture and infrastructure. Most importantly, I have the privilege of deep engagement with a community where caretaking the soul is seen not as a commercial or product-oriented enterprise, but one that is uniquely human.

Alison Vida

I was born in China and grew up in Portugal. Since high school, I was fascinated by the world of the deep psyche, which I discovered through books, and I would like to increase my knowledge in this field. I have chosen literature and languages as a platform of delight with poetry or literature, which portrays the inner world of personages and its fantasies. Also, an awareness of reality is well-described in so many romances and novels. Foreign languages were necessary to communicate and socialize with the broader others. Life took me to Switzerland where I met the Jungian world, a world where meaning and the depth of life was clearer. In my former profession as a teacher and translator, I had enough contacts with people and could be of use, but the capacity to empathize and talk about what really touches those around me was not actualized, which I felt I could do as a psychoanalyst.

ISAP stems from the hope and tenacity of some. It was the ideal place for a person like me. The friendly and competent environment of members of the faculty strongly, yet easily, guided my way through the school. On the contrary, the training is quite demanding and at ISAPZURICH this is known and I most appreciated the naturalness of human contacts in school and people's high ethical values. The extreme seriousness and respect in which patients are approached were factors that always have reassured me that this is the right school. Finally, the possibility to train completely in English and the flexibility of type of enrolment with its relative freedom to pursue major themes of interest were for me also decisive factors to opt for ISAP.

Nadia Yuan

A PROFILE OF JUNGIAN ANALYSIS:
A Position Paper

STEFAN BOËTHIUS & ISABELLE MEIER

In most cases, patients or clients have little idea of what a Jungian analysis is or its merits and uniqueness. We need to better communicate a clearly recognizable profile of Jungian analysis to these people. With this paper we want to make a contribution to the ways in which Jungians can explain Jungian analysis to people with no psychological knowledge. The challenge is to come up with formulations that explain our services in easily understandable terms.

To potential clients, two simple and readily understandable questions are:

1. Would you like medication, or do you want to talk?

This could be followed up with an explanation: "If you want pills, going to a psychiatrist is necessary. Psychiatrists are physicians who are permitted to prescribe medications. If you prefer to talk, then psychotherapy or psychoanalysis (or both) are more appropriate. There are cases in which people want to take medication and talk, but the basic principle is, if you don't want to take any pills, then you are right to come to us."

2. Do you want to talk about your symptoms or about yourself?

Psychotherapy often focuses on eliminating symptoms and establishing functional capability in the contexts of work and social relationships. Analysis has these goals too, but goes deeper since our work is based upon depth psychology. Fundamental problems, such as: "I want to find the real me" or "I no longer see any meaning in what I'm doing" are not amenable to a superficial symptomatic approach. To achieve sustainable healing in such cases, one needs to develop personality, discover of meaning, and improve the ability to bring one's life into harmony with one's own soul. A Jungian analysis helps individuals take responsibility for their own lives and provides greater consciousness and even transformation for oneself rather than mere treatment against symptoms. Often, in fact, symptoms themselves can further the process.

This explanation in response to question 2 is obviously not sophisticated enough to convey the essential aspects that distinguish Jungian analysis. To achieve this goal, we want to make a concrete comparison with the behavioral approach:

In the last ten or twenty years, the views of cognitive behavioral therapy (CBT) prevail in Switzerland and elsewhere. These determine what is meant by psychotherapy, how long a treatment should last, which procedures are appropriate, and what scientific criteria must be applied. Depth psychologists are collectively marginalized and generally regarded as supporters of an outdated method.

There's a saying: "If you can't beat them, join them." This would mean in effect to integrate CBT into the training of Jungian analysts. However, to "join" in this way has its pitfalls. Experience has shown that the stronger party determines the direction of the journey that the pack must follow. Specifically, this would mean that federal imposition of additional new training and certification requirements would deviate ever further from the Jungian point of view and take teaching with it. The profile of Jungian analysis would thus become less distinct.

If we do not want to be absorbed by the "mainstream" and lose our unique character, we must better define and articulate what we offer. The more clearly we can show the needs we meet, the better people will recognize the benefits of our approach. To be readily understood, we have selected five key differences between Jungian analysis and CBT (see Fig. 1).

In reality, Jungians may also apply behavioral and gestalt therapy or systemic elements in the course of a Jungian analysis. Analysis doesn't usually start with an immersion into the unconscious, but with practical problems, emotional knots, and psychological discomfort, which must be tackled first.

"Transformation" often presupposes reduction of suffering to a tolerable level and recovery of a certain capacity for functioning (i.e., attaining a degree of structure) through treatment.

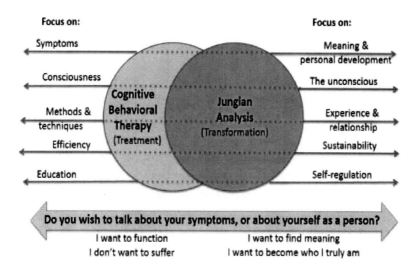

Fig. 1: Differences between a Jungian analysis and cognitive behavioral therapy

If this is attained, the analytical process is free to proceed in accordance with the following description by C.G. Jung: "... what the doctor then does is less a question of treatment than of developing the creative possibilities latent in the patient himself. What I have to say begins where the treatment leaves off and this development sets in."[1]

For Jung, this was the task of each and every individual—to seek their way to their true selves, to become more conscious, to develop greater authenticity, and to revive their own creativity. When the treatment phase is completed and the prerequisites for the transformation process are met, the aims of the analysis change to contrast significantly with those of behavioral therapy:

Instead of treating disease, the emphasis is on the *development of personality*. In this phase, accepting feelings like sadness, fear, anger, etc. is more important than alleviating symptoms by means of methods and techniques. These feelings are explored to provide a stimulus, or a wellspring, for the *search for meaning* and the process of *individuation* (becoming one's true self). Feelings should be endured, not repressed, because they are part of the human condition and may involve unconscious processes. It is not the symptom itself, but the meaning of the symptom, which is now at the center of personal growth.

Treatment in behavioral therapy consists of training adequate cognitions, such as attitudes, thoughts, views and beliefs, by means of questionnaires, inventories, pre-structured therapeutic approaches or even formulated manuals. Such treatment is systematic, focused on conscious processes. It seeks to reduce complexity and is generally based on a medical model. In Jungian analysis, however, individuals and their complexity, consisting of consciousness and the unconscious, are at the center; access to the unconscious is actively sought. There lies the creative potential for transformation of a person, especially in its undeveloped, repressed, denied, implicit, and unknown aspects. These unconscious contents can be explored and experienced through the interpretation and understanding of dreams, pictures, fantasies, fairy tales and myths, using Jungian concepts such as complex theory, the theory of archetypes, symbol theory, and the collective unconscious.

Behavioral therapy is based on methods, techniques and manuals that the behavioral therapist uses in dealing with the patient. The *relationship* with the patient plays a subordinate role, the person remains outside. Out of the debates among the psychoanalytic and Jungian schools over the last ten years the approach of *intersubjectivity* has evolved a common response to the challenge of the behavioral approach, emphasizing the therapeutic relationship between analyst and client. Jung had pointed out in his early writings that analysis is a unique process, an art[2], and not a technical procedure.[3] Jung said that in therapy "... the whole being of the doctor as well as that of his patient plays its part,"[4] and "Doctor and patient thus find themselves in a relationship based on mutual unconsciousness."[5] Jungian analysts today focus on these intersubjective processes and regard what happens between analyst and client as co-construction, as a "third," and as a process of "taking turns," which reinforces the clients' sense of their own being.[6,7] For this reason, personal *self-experience* is a central part of Jungian analysts' training, but not of behavioral therapists' training that gives greater weight to impersonal theory. Through the confrontation with one's own unconscious and personal development in the process of individuation, one's biographical background, strengths and weaknesses are identified and disruptive blind spots can be eliminated. A relationship that provides a secure framework is created, wherein painful material can be explored. A patient who can speak openly about himself has better chances of relieving his symptoms and improving his quality of life.[8]

A behavioral therapy session is usually shorter than a Jungian analysis. It is regarded as an efficient form of therapy that quickly makes an impact and minimizes the symptoms through a conscious configuration of the processes of perception. Behavioral therapists focus on inadequate cognitions and hold that our thoughts affect our feelings and behavior. The objective of a Jungian analysis, on the other hand, with its depth-psychological approach, is to achieve a lasting structural change and bring about the individuation of the person. It is, in other words, a process of mental reconstruction, which takes time. A study of Jungian practice conducted in 2006 showed an average number of 90 sessions for a Jungian analysis (approximately 100 for women, and 70 for men).[9] The advantage of structural change is that patients learn to promote their own development and thus have better chances to make their lives more meaningful and fulfilling through self-contact. The *sustainability* of analysis is also reflected in patients who are better able to cope on their own with new psychological stress and life crises. The Jungian approach does not aim for a short and inexpensive treatment, but at sustainability and long-term effectiveness. As the Jungian Christian Roesler notes in his review of studies of Jungian analysis, Jungian research has shown significant improvements not only at the levels of symptoms and interpersonal problems but also at the level of personality structure. These improvements were maintained some six years after completion of therapy, or even longer, and some studies even showed further improvements in clients.[10]

Furthermore, our view on psychic disorders focuses not on the pathology, but on the healthy capacities of the person. If the *self-regulating dynamics of the psyche* that the Jungian approach basically posits is interrupted and blocked by the problem, it must be for a good reason. A famous quote from C.G. Jung on the subject of neurosis shows his point of view in this context: "We should not try to 'get rid' of a neurosis, but rather to experience what it means, what it has to teach, what its purpose is ... *We do not cure it – it cures us.*"[11] We too keep recovery of performance in mind, but our further goal is the awakening of the creative potential and promotion of individuation (becoming one's true self).

NOTES

[1] C.G. Jung, "The Aims of Psychotherapy" (1929), in *The Practice of Psychotherapy. The Collected Works of C.G. Jung*, vol. 16, ed. Sir Herbert Read, Michael Fordham, Gerhard Adler, William McGuire, trans. R.F.C. Hull (Princeton, NJ: Princeton University Press, 1963), § 82f. (Further references to Jung's *Collected Works* are listed with chapter titles followed by volume and paragraph numbers.)

[2] C.G. Jung, "The Problems of Modern Psychotherapy" (1929), CW 16, §153.

[3] C.G. Jung, "Fundamentals of Practical Psychotherapy" (1935), CW 16, §6.

[4] C.G. Jung, "The Problems of Modern Psychotherapy" (1929), CW 16, §163.

[5] C.G. Jung, "Psychology of the Transference" (1946), CW 16, §364.

[6] Joe Cambray and Linda Carter, "Analytic Methods Revisited" (2004), in *Analytical Psychology: Contemporary Perspectives in Jungian Analysis,* ed.

Joe Cambray and Linda Carter (Hove: Brunner-Routledge, 2006), pp.116-148.

[7] Jean Knox, "Selbstwirksamkeit in Beziehungen – ein interpersoneller Blick auf das Selbst," in *Analytische Psychologie* 170, (April 2012): pp. 450-471.

[8] Sigried Schwandt, Thorsten Jakobsen, "Analyse der Therapeutentexte – ein Versuch der Quantifizierung qualitativer Aussagen," in *Seele und Forschung. Ein Brückenschlag in der Psychotherapie*, ed. Guido Mattanza, Isabelle Meier, Mario Schlegel (Freiburg: Karger, 2006), p.126.

[9] Guido Mattanza, Thorsten Jakobsen, Jacqueline Hurt, "Die Ergebnisse der Version Schweiz der Praxisstudie Analytische Langzeittherapie," in *Seele und Forschung: Ein Brückenschlag in der Psychotherapie,* ed. Guido Mattanza, Isabelle Meier, Mario Schlegel (Freiburg: Karger, 2006), p.75.

[10] Christian Roesler, "Evidence for the Effectiveness of Jungian Psychotherapy: A Review of Empirical Studies," in *Behavioral Sciences*, (March 2013): pp.562–575.

[11] C.G. Jung, "The Current State of Psychotherapy" (1934), CW 10, §361.

THE ROLE OF ISAPZURICH IN TODAY'S SOCIETY

BERNHARD SARTORIUS

The assumption that one of our main objectives in Western societies is to aim for "good functioning" is questioned less and less. This applies to the workplace and education, our private lives, and our important relationships (i.e., couples, parents-children). It has become increasingly clear that "normality" is equated with "good functioning," which can be measured against increasingly stringent "benchmarks" with correspondingly "efficient" methods. The development of this underlying notion of normality, since it has been taken over by politics and official bodies, seems to have acquired a quasi-totalitarian status. Anything that deviates from this good functioning is relegated with official and administrative sanctions as "pathological" or "antisocial." For example, the new DSM-5 (the Diagnostic and Statistical Manual of Mental Disorders, 5th edition: American Psychiatric Association, 2013) includes more and more "diseases" which hitherto were part and parcel of normal life, but are now classified as "pathological" behavior and feelings. For example, mourning after bereavement—if it lasts longer than 6 months—is defined as depression.

Needless to say, this development has generated ongoing costs in the billions, together with corresponding profits, on the one hand for psychopharmaceutical companies and on the other for a constantly expanding range of psychotherapeutic services. Thus, it is apparent that our contemporary concept of good functioning is closely linked to, and most likely has its origin in, the dominant scientific-technological and economic pattern of thought. The rarely questioned collective goal of endless economic growth, coupled with a just as rarely questioned exponential expansion of technical development, objectively coerces every individual to unceasingly accelerate and improve their work performance, their accumulation of knowledge, and the management of their personal problems—in relationships, in acquiring a sense of meaning in life, and in perpetuating good health. It is understandable that, under these circumstances, psychiatrists and psychotherapists try as much as possible to meet this objective, especially since they are directly exposed to the same pressure for efficiency by the health insurance system. Thus psycho-pharmaceuticals and suggestion-based psychotherapies promise the fastest and cheapest ways to combat psychological symptoms.

In my view, the pressure of this increasingly one-sided model of normality and good functioning continues to overburden our contemporaries, pushing them to their breaking point (burn-out) and paralleling the pressure that our civilization exerts on nature. So it is inevitable that—as is usual with any one-sidedness—compensatory movements and ideas develop and are being rediscovered. In parallel with green tendencies in politics, there are efforts to consider mental suffering not only as disturbance of normality and efficiency. As my colleagues writing in this *Festschrift* show, the psyche is infinitely greater and infinitely more differentiated than the aforementioned concept of efficient normality and its associated objective of combating suffering.

This new—and also very old—view is linked to a fundamental understanding of humankind in which we experience depths and heights, good and bad fortune, love and hate, faith and revolt, pain and happiness, beauty and destruction, etc. primarily as *ensouled beings*. This new perspective could form an oasis in the middle of the desert of our rational-technological society, be the germ that engenders the continuing expansion of consciousness: expansion insofar as the current technologically and economically limited of view of man and nature will soon come up against its own limits anyway.

More than 60 years ago, Carl Gustav Jung, after some initial hesitation—hesitation because he was still a child at the end of the 19th Century with all its faith in progress—realized that neither the individual nor society could really move forward with a purely rational, technological and economic point of view. He wrote:

> ... I know from my work with my patients, as well as pupils, how much the modern mind is in need of some guidance and how helpless people are in

envisaging and dealing with the enormities the present time and still more the immediate future will present us with. I cannot help believing that the real problem will be from now on until a dim future a psychological one. The soul is father and mother of all the apparently unanswerable difficulties that are building themselves up into the heavens before our eyes. We are thoroughly in need of a new orientation...[1]

In other words, the psyche is infinitely more comprehensive than the immediate needs of the ego for more happiness and less suffering, and it attempts to manage these needs in a techno-rational way. The night dreams of every individual, together with art, literature, and religion, all indicate that there are psychological openings to a profound experience and understanding of individual and collective psychic life.

In *this* sense we try, with the Jungian analyst training offered at ISAPZ-URICH, to make a contribution to psychotherapy training within our modest limits, which attempts to do justice to the term "psychotherapy" ("healing the soul"). We train people who are willing to embark on the adventure, which entails a real openness to the psyche and who later accompany their "patients" in facing this challenge. It is not our aim to obstruct the aforementioned current trend for efficient and rational management of mental suffering and its corresponding methods. What we are trying to do, which cannot be viewed through the lens of techno-scientific hermeneutics and their objectives—because it is for them *invisible*—is to divine the qualities of the soul and to connect them with the reality of life and the psyche. When Paul Klee painted a village, his imagery did not contradict a surveyor's view of that village—the portrayal looks the same, but different.

NOTE

[1] C.G. Jung, "Letter to Werner Bruecher of 12 April 1959," in *Letters 2: 1951-1961* (London: Routledge & Keegan Paul, 1976), p. 498.

TEN YEARS OF THE ISAPZURICH PROGRAM COMMITTEE

NATHALIE BARATOFF, DIRECTOR OF PROGRAM

History

Ten years on seems a good time to pay homage to the past and take stock of our achievements. It is, moreover, a time to check our compass, to make sure we are still on course.

Reading through the minutes of our first meetings takes me back to a situation so different from, and yet so similar to, where we stand today. Chills go down my spine as I recall the exhilaration of a new beginning, the risks involved, and the need to plunge into new activities with little or no advance preparation. At that time we were held together by the common conviction in the necessity of our undertaking. Much of the emotional involvement has faded today, but to re-read the old documents that record our step by step development is to once again become immersed in the atmosphere of those times.

This set the stage for the Program Committee's (PC) first meeting on October 26, 2004, nearly two months after ISAP's founding. The original committee consisted of the following members:

Ian Baker	John Hill
Nathalie Baratoff	Ursula Lenz-Bücker
Maria Anna Bernasconi	Urs Mehlin
Susanne Boëthius	Cedrus Monte
Brigitte Egger	Bernard Sartorius

In the first semesters of our existence, we were faced with the challenge of making a new start without cutting ourselves off from our roots of instilling a new breath into a program, which remained basically the same as that of the Institute. This was necessary not only for the continuity of our training but also for ISAP as a whole for we had had no quarrel with the contents of the Institute's training program. Indeed, from the start ISAP has retained the bilingual program of analytic training with two fourteen-week semesters each year in Zürich. Today we are the only IAAP-recognized institute worldwide that adheres to a full immersion model—and one of the few that still adhere to Jung's interdisciplinary outlook by admitting students with advanced degrees in any field of study.

An initial but non-substantial change we did undertake was to restructure the semester program booklet according to subject areas with German and English courses side by side rather than in two separate sections. This change came with certain difficulties because, as we still see today, some courses don't fit easily into our subject areas. In general, however, we feel this layout affords an improved overview of the curriculum.

Program Overview

The interdisciplinary subject areas taught and examined today are basically the same as they were long before ISAP's founding:

- Fundamentals of Analytical Psychology • Psychology of Dreams
- Psychological Interpretation of Myths & Fairy Tales
- Psychological Interpretation of Pictures • Ethnology & Psychology
- Religion &Psychology • Association Experiment & Theory of Complexes
- Developmental & Child Psychology • Comparative Theories of Neurosis
- Psychopathology & Psychiatry • Individuation • Practical Case

It has always been the case that many analysts prepare their courses aiming to present them within the given study areas. These colleagues favor a structured program in which the available courses cover the material that will be subject to examination. Other analysts have focused foremost on a gripping idea, a topic of the moment. Such material, they concede, might or might not connect directly with the exam areas—but it does connect directly with the creative source, which we equally want to convey in training. Colleagues of this persuasion recognize that post-graduate students need not rely entirely on a semester program, for they are accustomed to independent work, including preparation for exams. Such was the attitude in the early days of the Zürich Institute—and I have indeed seen a semester program "booklet" from that time in which the complete courses listing fit on a single sheet of paper, albeit printed on both sides.

Always wanting both, the Program Committee schedules courses in all given subject areas and also courses that broaden the scope of training. Among the latter, we have been pleased to present topics, such as the following:

"The Alchemy of Training" by Deborah Egger

"The Politics of Complexes" by Lawrence R. Alschuler

"Talks and Films," a presentation of Peter Ammann's life's work

Screening of the film *Sabina Spielrein, Freud and Jung*, which was attended by its director, Roberto Faenza

"The Role of the Anima" by Pedro Kujawski, accompanied on the piano with excerpts from Schumann's work

Starting in the winter semester of 2006-2007, we added an annual series of five to six lectures to the regular program on a topic of general interest. The first of these Ring Lectures was dedicated to "Music and Psyche," and ended with a benefit concert of Russian folk songs. Then followed "Art and Psyche" in 2007, "Dream and Culture" in 2008, "Dance and Psyche" in 2009, "Saints and Sinners" in 2010, "Trauma and Psyche" in 2011, "Ritual and Psyche" in 2012, and "Synchronicity and the Unexpected" in 2013.

Further examples of special events are the theatre productions of *The Jung-White Letters* and *Scenes from the Red Book*, which are making ISAP known far beyond Switzerland's borders. The same can be said of the Jungian Odyssey and the Zurich Lecture Series. The Märztagung programs in German provide a much-needed balance to the English-language events.

Such events are by no means restricted to our students and participants. We intend them also as an outreach to the greater community of which we are a part, acquainting those outside ISAP with our activities and furthering the knowledge and appreciation of Jung's contributions to the field of psychology and beyond.

At the end of each spring semester, during Prelude Week, which precedes the Jungian Odyssey, the PC plans a semester excursion to a site connected with Jung's life and work. Among those already visited are his lakeside house home in Küsnacht, his self-built tower in Bollingen, the Psychological Club, (which originally housed the C.G. Jung Institute), and Burghölzli (the psychiatric hospital where Jung began his professional practice).

The Way We Work

Over the past ten years, two tools have come to greatly simplify our preparation of the semester program. The first, developed in 2009, is a seven-page workflow table that specifies every step of the way: the "who," "what," and "when" of all the big and little chores needed to ensure that students receive the new program before leaving for semester break. Although colleagues outside of the PC sometimes find it difficult to accept our seemingly rigid workflow, for us it has become an invaluable tool.

An equally helpful tool is the online course proposal form. This facilitates presenter registration and considerably reduces the committee task of gathering and processing proposals. Despite initial difficulties with the digital format, most colleagues are now sufficiently acquainted with this tool to submit their proposals online. The new procedure has greatly simplified and streamlined our work.

Since the activation of subject area departments in 2008, the PC has greatly profited from their support in supplying the necessary courses for each semester's program. We hope that this will continue because, after all, the department members hold the expertise to insure that their respective subject areas remain up-to-date and properly represented in the semester program.

From the very start, Stacy Wirth has worked behind the scenes, helping us with the layout of program material. Prior to the introduction of the electronic course proposal forms, she drew up the registration forms for proposals; now her touch can be seen in all ISAP flyers. The PC has always appreciated her creative contribution to the semester program.

Urs Mehlin, a co-founding member of the Program Committee, took over as the chair from 2009-2011. At this time the workflow table was first put into use, and the final scheduling of the program was outsourced to Shona Boisvert, who has since remained a reliable assistant.

Most deserving of thanks are all the ISAP analysts who, semester after semester, contribute the courses that make up the program, for clearly without their contributions there would be no program to talk about. We are also most grateful to colleagues who supply us with suggestions for guest speakers, a number of whom return to teach at ISAP so often they become long standing presenters.

It has long been a policy of the PC to introduce first-time lecturers to their audiences, as well as to invite them for a meal and, where possible, to provide them with lodging. Such contacts have always been interesting and enriching, enabling us to become better acquainted with colleagues outside our school and to learn more about their work.

Thanks to all of the above measures, we have become better organized and more focused, and we have now been able to reduce the number of people required to carry out the committee tasks. At the end of Fall Semester 2013 the PC was composed of:

Nathalie Baratoff	Nancy Krieger
Irene Berkenbusch	Andrew Fellows
Maria Anna Bernasconi	Ann Chia-Yi Li

We also wish to mention the greatly appreciated student representatives who connect us with the student body, acquaint us with its concerns and needs, and provide invaluable feedback.

Our Guiding Tenets

It has always been the PC's prime concern to provide our students with a solid foundation in the theoretical and practical aspects of analytical psychology. Courses offered in the semester program support and supplement the work done in analysis and supervision; moreover, they introduce students to different approaches. For example, individual analysts may have different ways of working with dreams; some work with pictures, some put emphasis on transference, some on fairy tales, some on body work, some are more clinical in their orientation, others more archetypal, and so on.

ISAP is privileged to be able to offer a program that is rich in diversity and broad in knowledge and experience. This is made possible by the large number of participating analysts and guest lecturers coming from many different countries, cultures, and academic backgrounds. But size and diversity are not our only goals: We also feel responsible for giving the semester program direction and shape.

The recent introduction of Swiss federal law governing training in psychotherapy made us all acutely aware of the need to consider the long and short term consequences for ISAP. The PC has contributed to the process by examining the federal accreditation requirements with regard to curriculum and showing that their integration in our program appears to be possible.

Yet we remain firmly convinced that compliance with the law cannot come at the expense of ISAP's core values. In particular, we are unwilling to jeopardize students' central encounter with the reality of the psyche[1] for in a Jungian perspective, this is the "the *sine qua non* of all existence."[2] (Jung's italics) It is responsible for all those "miracles" that befall us in analysis, all those synchronicities and solutions that suddenly emerge as if from nowhere when all conscious means are exhausted. Moreover, it is the very experience of the reality of the psyche that changes and deepens our students' lives as they go through the training.

It is through our dreams that we are usually introduced to this reality, but in training, students come to recognize it in pictures, myths, fairy tales, in day-to-day fantasies, emotions, projections, and relationships. It seeps through in their studies of ethnology, religion, the complexes, transference and counter-transference, and even psychopathology. Students learn about the reality of the psyche from many sources but, ultimately, they will only be convinced by direct and personal experience.

Such a process requires time and openness toward the irrational. The Zurich model has always viewed the Jungian analyst's profession as a calling, an initiation into the mysteries of the psyche. It requires incubation, a voluntary exposure to the unconscious and a giving up of any preconceived plan of action. In ISAP's yearly 28 weeks of residency, the student has the opportunity to develop according to his or her individual needs, to choose whatever courses are relevant at the moment and to integrate the material gleaned from these courses into personal analysis and work with clients.

The profession of Jungian analysts is not only a calling, it is also an art and, as such, it requires a steady and uninterrupted process of learning. It is for this reason that ISAP insists upon a training program in which analysis, curriculum, case work, and supervision are ongoing and take place concurrently in time and space.

The Greeks called it *temenos*, the alchemists referred to it as the hermetic vessel. Both illustrate the process of letting things happen, both require a closed space where that "other" part of the psyche can be heard and responded to. Such are the considerations that guide the PC in shaping the semester program, for Jung's words ring just as true now as they did when spoken in 1957: "...*the world today hangs by a thin thread, and that thread is the psyche of man*"[3] (original italics).

NOTES

[1] C.G. Jung, "The Real and the Surreal" (1933) and "The Stages of Life" (1930-1931), in *Structure and Dynamics of the Psyche: The Collected Works of C.G. Jung*, vol. 8, ed. Sir Herbert Read, Michael Fordham, Gerhard Adler, William McGuire, trans. R.F.C. Hull (Princeton, NJ: Princeton University Press, 1963). (Further references to Jung's *Collected Works* are listed with chapter titles followed by volume and paragraph numbers.)

[2] C.G. Jung, "Psychological Commentary on the Tibetan Book of the Great Liberation" (1939), CW 11, §769.

[3] C.G. Jung, "The Houston Films [1957] with Richard I. Evans, 'Jung on Elementary Psychology (New York, 1976),'" in *C.G. Jung Speaking: Interviews and Encounters*, ed. William McGuire and R.F.C. Hull, Bollingen Series XCVII (Princeton NJ: Princeton University Press, 1957), p. 303.

WHEN ISAPZURICH WAS ESTABLISHED
The Presidency of Paul Brutsche 2004-2008

ISABELLE MEIER (IM) INTERVIEWS PAUL BRUTSCHE (PB)
at his office in December 2013

Prior History: September 2003 to October 2004

IM: Could you briefly recount how ISAP came into being and what preceded it?

PB: ISAPZURICH was set up in a relatively short time. Because of this, the Curatorium of the C.G. Jung Institute thought that our institution has been planned and set up well in advance, but that was not the case. We simply worked very intensively for four to five months to establish a new institute. We were forced to do so after we realized that the actions of our analysts against the then autocratic Curatorium were to no avail. In particular, the *pledge campaign* in which the co-signatories had agreed to pay a specified contribution to alleviate the Institute's financial crisis if the Curatorium would resign was ineffective. Also the action against the obligation to sign a so-called *membership declaration* introduced by the Curatorium did not produce a solution (see below). The eventful history is long and should not be further elaborated at this point, but the contributions of Debbie Egger and Stacy Wirth in this context should be acknowledged in this publication. I would just like to mention briefly, at this point, the history that led directly to the creation of ISAPZURICH.

On September 13, 2003 an open meeting of 88 Jungians took place at Erlengut in Erlenbach where it was advocated strongly that a solution to the crisis at the C.G. Jung Institute had to be found. We convened because the Institute had sent a letter on July 15, 2003 to the analysts that asked everyone to sign the aforementioned membership declaration, which affirmed unconditional loyalty to the Institute and willingness to provide financial support in the future. It also threatened that anyone who did not sign would not be eligible to take on new training analyses and supervision or participate in committees, beginning in the winter semester 2003 to 2004. This membership declaration caused a strong reaction among the analysts and subsequently led to the open meeting at which we agreed unanimously that the Curatorium had no right to demand a membership declaration from us and that we would therefore not enter into it.

As a result of that meeting, we began to think about creating a new training program in Zurich. A group of analysts was intensively involved in this process, along with all those who supported us financially or ideologically.

IM: But the intention wasn't initially to establish a new institute?

PB: No. At the open meeting we decided we wanted to contribute something to the preservation of the Institute, hence the pledge campaign. Unfortunately we had to realize again and again that this was impossible. The pledge campaign was launched on 29 April, 2004 by Nathalie Baratoff, Peter Ammann, Stefan Boëthius, myself, Deborah Egger, Jan Peter Hallmark, Franz-Xaver Jans-Scheidegger, Hans-Peter Kuhn and Monique Wulkan to keep the financially troubled C.G. Jung Institute going. Our treasurer, Stefan Boëthius, informed the Curatorium of the outcome of the initiative on May 17, 2004. The pledge campaign was worded as follows:

> To support the Institute, I pledge to contribute funds under the condition that the present Curatorium withdraws no later than the end of the summer semester 2004, and is replaced by a new Curatorium previously constituted according to the results of a consultative election held by the community of analysts. The sum of pledged contributions will be reported to the Curatorium and the analytical community (without naming the individual donors) after May 14th, 2004.... Accordingly: By Monday, May 17th, 2004 the pledged funds amounted to CHF 336,264.[1]

The Curatorium did not respond to our initiative. We had to realize that our various actions had not borne much fruit.

IM: What happened to the pledged money?

PB: A portion of the pledged money was donated later to ISAPZURICH. In October 2004, we had CHF 260,000 in donations that were paid to our school, although they couldn't be tax-deductible at the time. We only worked out tax exemption status for donations to ISAP later. One and a half years later—in February 2006—this sum had grown to CHF 520,000. [2]

IM: What happened next?

PB: On July 6, 2004, over 100 analysts gathered in the parish hall of St. Andrew's Church in Zurich with the intention to establish a new training institute. The name and location of the institute were not yet fixed at that time. It was merely a first information event. We then needed legitimacy and support from the AGAP members' meeting to embark, as a subgroup of AGAP analysts, upon an AGAP training program in Zurich. This meeting took place at the IAAP Congress from August 29 to September 3, 2004 in Barcelona, during which, as usual, AGAP held its business meeting. An amendment of AGAP's statutes, proposed in advance in writing, explicitly stated the pre-existing right of AGAP to operate its own training program.

IM: When and in what form did ISAP become a concrete reality?

PB: Immediately after the AGAP business meeting in Barcelona, the inaugural assembly for ISAPZURICH was held on September 9, 2004 at Erlengut in Erlenbach. It was not until this time that some very basic things were decided. The level of contributions for ISAPZURICH membership was divided into Category A (annual contribution of CHF 600) and Category B (annual contribution of CHF 1000). The latter was intended for those who did not actively participate, but wanted to support the seminar ideologically. Also the first official posts were determined and people were elected to these posts. In Erlenbach, 58 participants were present. Karin Evers and I stood as candidates for the presidency. The minutes of the meeting record that:

> The election of the President is strongly debated. Opinion divides along two basic lines: (1) One maintains that, with all due respect for Paul Brutsche, his presidency could be encumbered by the history of his involvement with the present Curatorium; Karen Evers would therefore be desirable. (2) The

other maintains that, in this difficult start-up phase, we urgently require a president who would assume office with experience; in this case, Paul Brutsche would be preferred.[3]

In the end, I was elected by 34 votes. Stacy Wirth was elected as Vice President, Stefan Boëthius as Treasurer, Karen Evers as Director of Administration, Nathalie Baratoff as Director of Program, Katharina Casanova as Director of Studies, and Doris Lier as Director of Admissions. We had assembled the Board.

The Establishment of ISAPZURICH in October 2004:
The name

IM: What led to the rather cryptic name ISAPZURICH (Internationales Seminar für Analytische Psychologie [International School of Analytical Psychology])?

PB: That was a long and complicated process. The name should of course indicate clearly that our field was Jungian analysis, yet the word Jung could not appear directly. It was necessary to distinguish ourselves from the name of the CGJI to avoid legal actions. After we had found the name ISAPZURICH, we checked on legal grounds whether there were other institutions with similar names. It turned out that we were the only one. Indeed, the Israeli group (Israel Society of Analytical Psychology: ISAP) came dangerously close but could not be embellished with the addition ZURICH.

IM: How happy were you with the outcome of your obviously time-consuming name search?

PB: We had to admit to ourselves that we could not be very happy with our name. Being able to use the name C.G. Jung Institute was a great asset that we had to do without. However, in our name we emphasize the international aspect that is so important for us. Another challenge was to find a name whose acronym was the same in German and English. We were able to find this with "International School" and "Internationales Seminar" just in time for the launch of the first semester in October 2004.

Our first premises at Hochstrasse 38 in Zürich

IM: ISAP needed a home. How did you find it?

PB: We started to look for suitable premises in the summer of 2004. This was a difficult task. It should be accommodations that met our needs, not too small, not too big, not too expensive, reasonably accessible, and with a welcoming atmosphere. At that time I coveted every building in Zurich in terms of whether it would be suitable for ISAP. John Hill knew of a detached house in close proximity to the Epiklinik and visited it but received the disappointting news that the house would soon be demolished. Then Doris Lier casually told my wife and me in the car that she might have found a place, whose owner she knew well. It was the house on the Hochstrasse. This proved to be actually a stroke of luck, because Herr Maissen, the owner, was very generous to us in many respects. He installed a new floor, new toilets and new bookshelves, and compromised with the rent. He was willing to commit to us because, as a former founder of AKAD, he had sympathy for a new training enterprise, and because he was also open to Jungian psychology.

Fall opening, 2004

IM: What preparations were necessary in order to start the actual program and to bring students into the training?

PB: First we had to create training regulations. These were congruent with the training regulations of the Institute in most respects. We therefore labeled them as "provisional." We saw no reason to introduce new rules, as these had been tried and tested at the Institute. In addition, this continuity allowed recognition within the training requirements of the performance to date of students transferring from the Institute. During the first few years, however, we revised both the language and presentation of the regulations, as well as their content. In this final form, they were also published on our website.

Then there was also the matter of creating a course program. Nathalie Baratoff, together with the Program Committee, managed from the outset to put together a full program of lectures and seminars. The richness of the program, and the variety and quality of our presenters, was—and remains—an important factor in the appeal of ISAP. As before, ISAP could offer the most complete range of regular courses in analytical psychology worldwide.

Last but not least, in order to better structure the complicated training, with foreign and Swiss students, auditors, training and diploma candidates, German-speaking and English-speaking students, etc., and thus avoid too much chaos and to ensure some form of fair objectivity, we had to create guidelines and administrative forms en masse. Stacy Wirth and Doris Lier, who were responsible for the editorial and graphical implementation, did a masterful job in this regard.

IM: And then the first students came.

PB: Right. Originally we didn't know how many students we could expect.

We knew that we needed at least 20 students in order to make ends meet. In the end, we started with 37 students, which gave our Director of Studies, Katharina Casanova, a lot of of work. Among the students were 20 training candidates, 14 diploma candidates, and 3 matriculated auditors of whom 32 were English-speaking and 5 German-speaking. It seemed to be predominantly English-speaking students who came to us (see reference 2). We asked the attorney, Martin Amsler, to clarify whether foreign students could study at ISAP and would receive a residence permit. We needed to be recognized as a training institution by the immigration office. We had to explain who we were and what training we offered. Finally we received the green light from them. The C.G. Jung Institute already possessed this authorization. For us, this was important because without such recognition our foreign students would have received only tourist visas valid for 3 months, but no student visas.

Further legal (and other) hurdles

IM: With that, were all difficulties cleared out of the way?

PB: No, we had to overcome more hurdles. In addition to the legal clearance of our name and recognition by the Migration Office, we also needed the IAAP's confirmation that it recognized our diploma. Three members of the Institute had legally challenged the amendment of AGAP's statutes and the fundamental decision to create its own training program, which had been taken at the AGAP business meeting in Barcelona in the summer of 2004. Markus Wirth, our lawyer, defended the decisions, and AGAP finally won the legal case, thanks to his brilliant work (see the chapter by Debbie Egger & Stacy Wirth). The Curatorium also sent a letter to our students, claiming that it would be risky for them to study at ISAP. They alleged that it was not at all certain whether the diploma would be recognized for IAAP membership.[4] Christian Gaillard, the president of the IAAP at that time, finally confirmed in writing that, because our training qualified us for membership in AGAP, it also qualified us for membership in the IAAP.[5] The blessing of the IAAP was thereby guaranteed. Finally, we still had to fight for Charta membership to be recognized by the Canton of Zurich as a psychotherapeutic training institute. This was a very complicated process. To provide proof that our method and our training concept had high academic standards, many colleagues were actively involved. Indeed, the coordination and final editing of the resulting extensive text ended up in your hands, and you, Isabelle, did a great job. In September 2006, we were granted provisional membership, and a year later full membership.

ISAPZURICH in the first two semesters

IM: Could you describe what happened in the first semesters?

PB: We started out with our management meetings at a breakneck pace of around 30 sessions in the first year. The meetings were always held on Wednesday and regularly lasted late into the evening. Nathalie Baratoff once said that at no other time in her life did she hear the Swiss national anthem at the end of the day's radio transmission during her late-night ride home in the car as often as in those times. In the 2004-2005 annual report, I wrote:

> From the standpoint of the Officers Committee this first year of ISAP's existence was completely under the sign of start-up and development: The launching of this new training program has been by nature a very demanding and work-intensive job – if also an extraordinarily satisfying and motivating one, because so much is given back. In no time at all and out of noth-

ing, we were faced with the challenge of bringing forth an institution that met the complex requirements of any Jungian training program. In the course we were repeatedly confronted with the need for quick decisions in the greatest imaginable variety of areas. The consequent demand for constant movement and strength of resolve – admittedly not always part of the therapist's natural equipment – kept us constantly learning.[6]

This first annual report has been included in its entirety in the Appendix to convey an impression of all that had to be discussed and decided in the first year.

IM: In what ways was ISAP different from the CGJI?

PB: I would like to mention two points. The first relates to organization and the other to atmosphere. Our organization expanded substantially, as I already said, on the regulations and organizational structure of the C.G. Jung Institute with two important exceptions:

On one hand, we dropped two tasks from leadership responsibility, which had been under the control of the Curatorium at the Institute: the appointment of members to the various panels and committees and the appointment of teaching and supervision analysts. To do this, we created a Nominations Committee that submits proposals for new post-holders to the General Assembly to ensure that these are not, as had been the case, handed down from above, and thus adherence to democratic processes is paid enough attention. We also created a Promotions Committee, whose task is to interview aspiring training analysts and supervisors and to recommend their promotion to the General Assembly. These ideas were introduced by Debbie Egger to curtail the power of the leadership in two important areas and thus safeguard community spirit.

On the other hand, ISAP was structured democratically, according to the criteria of Swiss law on associations, so that important decisions are taken together at the annual General Assembly through various motions subject to regular voting. This scheme has created a fruitful spirit of joint responsibility and active participation among the analysts, which is largely responsible for the good atmosphere at ISAP.

As for the emotional aspect of the atmosphere, in the early days of ISAP I myself had a dream in which a dancing African woman appeared at an ISAP event. As president, I adopted the rhythm presented by the woman as well as I could, even though I had no reason to think of myself as an exceptionally talented dancer. Gradually the uplifting rhythm was transmitted to the whole community. This dream came as a great relief, especially given the background of the previous destructive atmosphere experienced at the Institute. It seemed to say that a positively invigorating, elemental, animating energy, which can again build something, something soulful, something creative, is once more possible. This is how new initiatives, such as the subsequent Jungian Odyssey, became possible. There were ideas and impulses from below, which were responded to with initiative and commitment. Thus unusual, creative events were realized, such as the singing event in the auditorium of Rämibühl Gymnasium with Nathalie Baratoff at which she sang wonderful Russian songs. All this was only possible because the energy sprang out from the community itself.

Maybe the newly introduced points system should be mentioned in this context. It was an innovative instrument to reflect the resources available at any given time by regulating remuneration for work undertaken. Under this scoring system, our treasurer, Stefan Boëthius, presents the profit (or loss) of the preceding year to each General Assembly in the spring, and we discuss together how we want to split the profit. A certain number of points are assigned for each activity and a value per point is agreed upon. In the discussion, the value of points can be increased, risking a loss, or reduced to save money.

Administration, office furniture, copier etc.

IM: How did you actually build up the infrastructure of ISAP as needed for its administration? And how did the staffing develop?

PB: At the beginning there was only one phone in the office. Urs Mehlin then generously donated a TV set, a VCR, and two overhead projectors. I had discovered an internet-order company for office supplies with the symbolic name *Viking,* where we could buy a fax machine, office chairs, and much more, cheaply. Slowly the necessary infrastructure came together. In our heroic early days we had no administration staff as such. We wrote everything and made copies ourselves. We worked so hard, and at such a pace, that we soon began to ask ourselves how long we could go on like this. At this point we decided to hire staff. In our Annual Report of that year, our Director of Administration, Karen Evers, wrote:

> ISAP opened its doors in October 2004 and immediately began a full schedule of classes and seminars. The organization of the front office and the logistics of managing the classes at that time was all done by volunteers. Despite the enthusiasm and high level of competence of the student and analyst volunteers, it quickly became apparent we needed a professional secretariat. The decision was made to hire an experienced secretary for a 60% position to supplement the 10% position filled by the admissions/director of studies secretary. We were very fortunate that Elena Eckels (whose long experience is invaluable) accepted to take charge of the admissions secretarial work and to find another very experienced and capable person, Franziska McSorely, who took up her position in March 2005. After the end of the summer semester 2005, it was again apparent that as the student body grew and the demands of the secretariat grew, we needed more capacity. The Officers Committee decided, therefore, to hire Karen Evers for a 20% position to work in the secretariat (due to the fact that she did not need any training and would be able to jump in immediately). This provides the secretariat an 80% position.[7]

We had discussions about whether a member of the seminar leadership could also be paid for a part-time job, as the seminar leadership actually worked free of charge. But as we had more students than planned, this generated a lot of additional work for the administration, which was led by Karen Evers. We had no choice but to overlook our principles in this case.

IM: How did ISAP's financial situation develop?

PB: Very well. In the first Annual Report, our treasurer, Stefan Boëthius, reported on our financial position, which he presented in very positive terms:

Thanks to the selfless engagement of many colleagues – and to the unexpectedly high rate of student enrollment – ISAP's first fiscal year closed as per 30 Sept. 05 with a bank balance of CHF 714,392.10. CHF 165,000.00 was the original sum budgeted. This is a healthy financial basis for ISAP's build-up phase. The sum of donations for this fiscal year was CHF 519,656.00. Expenditures reached a total of CHF 171,513.02, while the original amount budgeted was CHF 163,800.00.[8]

Gradually we built up our personnel. However, Stefan always had the high cost of administration at the Jung Institute in mind, which he considered to be the main reason for the financial crisis of the Institute at that time. It was therefore his declared intention that we should only increase staffing levels with extreme caution. However, we could also not ignore the fact that we were a complicated training organization that simply could not cope without more staff.

ISAPZURICH 2005 to 2006

IM: Could you tell us something about the next two years?

PB: We were able to recruit Elena Eckels to the secretariat. Many people still knew of her as Studies Secretary at the C.G. Jung Institute. We were delighted to have her with us to be able to draw on her extensive experience. We also started to build up our own library. The results of our appeal for voluntary donation of books to our library initially fell short of our expectations. We listed the books that we needed for our reading lists in the various subject areas on our website. With time, the list was getting shorter as colleagues gradually donated the books we were looking for. Finally we reached our target number of books, and our library now has about 4,000 items of which English-language books make up the majority. In response to our request, the Parrotia Foundation gave us a generous contribution, amounting to CHF 132,000 in three donations, which we could use to build up the library further and hire a librarian. To our great delight, Helga Kopecky also finally came to us from the Institute. She brought her immense expertise, including how to set up a library and catalog its contents with appropriate software.

We built our institute up further. Added to this was the collaboration with the IMD Business School in Lausanne. Jack Wood, a professor at IMD and diploma candidate at the Jung Institute Zurich, offered diploma candidates at ISAP the opportunity to work analytically with IMD MBA students. This provided a welcome opportunity for our diploma candidates to gain practical experience that counted towards the required hours of analysis with their own analysands.

On November 11, 2005, the Zurich District Court decided in AGAP's favor in its court case, thereby confirming the overwhelming majority vote of AGAP's members for the delegation of the training rights of AGAP to a Zürich subgroup. This confirmed ISAP's permanent standing in AGAP and the IAAP, and thus assured the international recognition of our training. We could breathe a sigh of relief.

In 2005, we were also admitted to the Swiss Charter for Psychotherapy. With this, another hurdle was overcome. In the same year we embarked upon our first Jungian Odyssey, a week-long retreat for members of the public interested in Jungian thought that also made ISAP known internationally (see the Jungian Odyssey chapter).

Unfortunately Franziska McSorely announced her resignation in March 2006, and likewise Karen Evers in October of the same year because she immigrated with her family to Canada. After this we separated the front and back office functions: in the future the front office would deal with phone and email enquiries, while the back office dealt with student enrollment, liaison with the Migration Office and other authorities. After the departure of Franziska McSorely, we were fortunate to recruit Karin Buchser, a competent and experienced secretary, for 70% employment. She brought the secretariat up to professional standards. Later she was appointed to Director of Operations and was entitled, as a non-elected member of the ISAP leadership, to meet with them to take strategic decisions about ISAP's organization, in addition to her administrative management role. We enlarged our available space by additionally renting the second floor, which we needed for the back office, two therapy rooms for the students, and a library storage room.

In February 2006, we already had 91 analysts who were participants of ISAP and 62 students (see Note 2). The hustle and bustle of the start-up period subsided a bit, but we were not finished with the construction of ISAPZURICH.

A request was submitted to the Swiss tax authorities to relieve ISAPZURICH from tax liability, and to ensure that donations for ISAPZURICH could be tax-deductible. Stefan Boëthius finally received confirmation that donations from Switzerland would be tax-deductible.

ISAPZURICH 2006-2007

IM: What can you say about the last two years of your presidency?

PB: By the fall semester of 2006 we were up to 82 students. English was the most widely spoken language for 57 students. Most students came from the U.S. and Canada, followed by those from Switzerland and Europe. However, more and more Asian students came to our institute, while the number from the U.S. and Canada declined slightly over the coming years.

In November 2006, Sandy Schnekenburger became the new Director of Administration. In 2007, Myrta Blarer joined the staff to look after accounting. Later, her employment was extended to 40% in the front office and 20% in the accounting department. This made it possible for our front office to also be staffed on Fridays from then on.

In 2007, we also introduced the post of ombudsman. This post was shared by two ombudsmen, whose function was to provide a focal point for, and advice and assistance to, analysts and students in the event of conflicts.

Finally, ISAPZURICH's continuation as AGAP's delegated training program was confirmed with a clear majority at the AGAP assembly held at the IAAP Congress in Cape Town in 2007. After this endorsement we could hope that the question of whether ISAP should continue to exist or not would no longer be raised in the future.

Having originally planned only one year as President, I resigned after three and a half intense and rich years. The dedicated and competent teamwork among members of the seminar's leadership and the tangible support from the analyst community had enabled ISAP as an institution to be created in a short time, which impressed and attracted our students and was recognized by our participants. After the years of the organizational ordering of the seminar's structure and the establishment of our program, I was happy to hand over the reins of office to the incoming co-presidents, Ursula Ulmer and Murray Stein, who were both competent and experienced in leadership.

NOTES

[1] See p. 191, Appendix Source No. 1: Initiative Pledge Campaign: Results, May 17, 2004 (email).

[2] See p. 193, Appendix Source No. 2: Development ISAPZURICH October 2004–February 2006.

[3] See p. 195, Appendix Source No. 3: AGAP-Meeting AGAP Meeting for the Constitution of the Delegated Training Program in Analytical Psychology, Erlenbach/Zürich at Erlengut, September 9, 2004, §4 (founding document).

[4] See p. 201, Appendix Source No. 4: Letter to the Curatorium, April 3, 2005 (email).

[5] See p. 204, Appendix Source No. 5: Letter: IAAP President Christian Gaillard to Mr. John Betts, September 13, 2005.

[6] See p. 205, Appendix Source No. 6: ISAP Annual Report, 2004-2005, §1 Officers Committee, November 2005.

[7] Ibid., §2, Administration, p. 205.

[8] Ibid., §7, Finances, Donations, Advertising, PR, Internet, p. 211.

STEERING ISAP ONTO THE HIGH SEAS

ISABELLE MEIER (IM) INTERVIEWS MURRAY STEIN (MS)
at his office in Zurich on February 13, 2014

IM: When you took office as president of ISAP in 2008, what was the situation and how did you see your task ahead?

MS: Actually, I took the job reluctantly. I had moved to Switzerland in 2003 during my term as president of the IAAP. After 2004, when the IAAP job ended with the Congress in Barcelona and the formation of ISAP as the training program for AGAP, I looked forward to having more time for teaching, writing, traveling, and doing things not connected to administration of organizations. Paul Brutsche was president of ISAP and had done a Herculean labor of creating the organizational structures that made it into the quality training program that continues today and was very nicely guiding the ship he had built on its maiden voyage. However, he was becoming exhausted after years of conflict at the old institute in Küsnacht and the huge creative effort required to bring ISAP into being and to preside over its first years of existence. Paul needed a break and wanted to hand over the presidency to other hands, and so he asked me if I would consider taking it over from him. As flattered as I was to be asked to do this, I was reluctant because I knew the requirements inflicted upon someone who took a responsibility like this. So at first, I demurred. But then I saw the need and agreed to take over the helm of the ship, provided a co-president could be found to assist me.

Fortunately, Ursula Ulmer was moving back to Zurich after living in South Africa for several years, and she agreed to step in as co-president. We served together for one year, and then I took the presidency while she assumed the role of vice president for a period. We worked very well together, and she lightened the load a great deal until I could find my feet and felt I could then carry the role by myself.

What I inherited from Paul was a carefully structured, very well organized training program. He built the ship—together with many others, of course—and it was very well done. It was modeled on the pattern of the C.G. Jung Institute as it had been during his term as president of the old Curatorium. It was a model with which I was very comfortable, which I believed in, and which I am still very happy to work with. I also found a group of analysts in ISAP who were very enthusiastic, with a wonderful spirit of community and cooperation, and they were willing to pitch in and do everything that was needed to maintain the excellence of the program. To me, the general atmosphere at ISAP felt very positive.

In the four years that I served as president, there were no serious conflicts among the analysts. The dues-paying participants taught and served happily on committees for "points" (modest tokens in monetary terms), and this made the school financially viable. I have been in other Jungian institutes and have observed many more, and often there were conflicts, old disputes, and difficulties that created unpleasant working conditions—that I didn't find at ISAP in the least. Maybe this has to do with the circumstances of its recent creation, I don't know.

It was in this spirit that I suggested to Paul Brutsche and John Hill already in 2006 that we create some theater at ISAP, staging a performance of the recently published *Jung-White Letters*. It was just an idea and came from a performance of the *Jung-Freud Letters* that had been staged in Chicago in the 1980s. My thought was that it would be suitable for the upcoming, jointly sponsored IAAP/IAJS conference to be held at the ETH in Zurich in July 2008. This would be a major staging place to show ISAP to the international Jungian community.

IM: I remember the scenes of the *Jung-White Letters* very well. I was at that time Co-Chair of the Jungian Odyssey, and we invited you to the Jungian Odyssey at the Waldhaus in Sils-Maria. There were many people present, including the director of the Waldhaus.

MS: Yes, I remember the scene very well too. Paul read the letters of Jung, and John Hill spoke for Victor White. They were very convincing. For that performance, Ann Lammers could not be present so we enlisted Dariane

Pictet to play the role of the Narrator. She did an excellent job and later also played the role of the beautiful *Soror mystica* at a performance in Braga, Portugal. We gave our first performance, with Heike Weis as *Soror mystica*, at the AGAP Forum in the classic Grossmunster Helferei auditorium in June 2007, and this was followed by the performance at the ETH in July 2008. Earlier that year, we had travelled to London and performed in a London theater for British colleagues and their friends. Then, in 2009 we performed at the Jungian Odyssey in Sils-Maria, and later that year again for the AGAP Forum at the Widder Hotel in Zurich, where the performance was filmed. Thanks to the Asheville Jung Center, a DVD was made of this performance and is now available for sale along with extras including interviews with each of the performers speaking about their experience playing the roles.

We gave our final performance in Montreal at the IAAP Congress in 2010, and it was here that Heike Weis met her future husband. Everywhere we received very warm applause indeed from our friendly audiences. I was confident that this would arouse interest in ISAP and gain us some recognition, and it certainly did. Paul and John have been often addressed as "C.G." and "Victor" when they travel abroad, and Heike Weis met her life partner as a result of her stellar presence on stage.

IM: What does it mean exactly, taking ISAP out onto the high seas and into the international communities?

MS: My particular mission as president, as I saw it, was to make ISAP better known to the international Jungian community, to take the ship that Paul had built now out onto the high seas, to fly the flag of ISAP in many ports around the world. I felt that my challenge in inheriting this well-built ship and guiding it ahead was not only to keep it afloat. I felt my challenge was to take it on the high seas, out into the international communities, out in the world—that was my vision.

When I became president, ISAP was not so very well-known beyond these shores. It was new and so, of course, had not had enough time to establish an international brand name. People had not yet heard of ISAP and didn't know what these letters stood for, even if many did know the familiar names of the senior training analysts at ISAP, like Mario Jacoby, John Hill, Paul Brutsche, etc. I saw my mission was to make ISAP known to the international Jungian community, to the IAAP membership worldwide, and I could do that because I had many connections from my twenty years on the IAAP Executive Committee. With the help of many others, I took a number of initiatives to make ISAP known. One was at the congress in Montreal in 2010, where AGAP sponsored a large party to introduce ISAP to the Congress. At that gathering I said boldly that ISAP offers the best training program in the world. This was of course "hype" and overstatement, American style. I wanted to sell ISAP to the audience, bring ISAP up to the "glamour level" of the old C.G. Jung Institute. I think it deserves to have it, because most of the best-known training analysts came over into ISAP when it was formed. That was what I wanted to tell to the audience. I spoke about these senior training analysts, famous names like Mario Jacoby, John Hill and Helmut Barz, who were very well-known internationally. I said we are running a training program that is unique in the world, nobody else in the world is doing a full-time international training program and with so many great analysts teaching and serving on the committees. The audience reaction was mixed, to say the least. I had waved a red flag in front of the competition and there was some criticism of my statements, but I felt it was worth the attacks. ISAP had to become known, and this helped raise the brand into collective awareness

IM: Didn't you also try to bring Jungians to Zurich?

MS: Yes, after all Zurich is the home of Jungian psychology. It seemed important for ISAP to remind Jungians worldwide of this obvious historical fact. I was therefore extremely pleased that the IAAP and IAJS, the two largest international Jungian organizations, agreed to hold a joint conference in 2008 at the ETH, where Jung was professor from 1933 to 1941 and where the Jung archives are housed. ISAP was very involved in this international and interdisciplinary conference, which took place July 3-5, 2008, with the title, "Contemporary Symbols of Personal, Cultural, and National Identity: Historical and Psychological Perspectives." The aim of the conference was to explore the many and diverse ways through which Jungian studies and analytic practice can contribute to today's urgent debates about the role and impact of myths, symbols and powerful narratives in culture, national identity, and politics as well as personal life. Debbie Egger, Allan Guggenbühl, Marianne Müller, Reinhard Nesper (Chair), and I made up the organizing committee. This was an effort on our part to show that Zurich is still an important center for Jungian scholarship and training. ISAP hosted an afternoon tea at its premises, and many participants at the conference came by for a visit. At that conference, too, I met Nancy Cater, and she and I came up with the idea to found the Zurich Lecture Series. This was a further attempt to attract people to Zurich and to showcase ISAP analysts.

IM: The Zurich Lectures Series is another chapter in this book, but can you describe how the connection with Nancy began?

MS: Nancy Cater had a longstanding love of Switzerland and wanted to make some further connections to the country and to Jungians who live and work here, hence her interest in ISAP. In a casual conversation, she and I began to develop the idea for a lecture series to be held annually in Zurich. It was modeled on the Fay Lectures Series in Texas and would feature Jungian thinkers from around the world coming to Zurich, writing a book for the occasion, and delivering lectures here over a weekend based on their new book. Nancy's company, Spring Journal Books, would be the publisher, and ISAP would host the event. This idea was also derived from the famous Terry Lectures at Yale University, which Jung delivered in 1936 ("Psychology and Religion"). In partnership with Spring Journal Books, ISAP launched the first of what would be known as the Zurich Lecture Series in October 2009 with ISAP training analyst John Hill brilliantly speaking on the theme "At Home in the World: Sounds and Symmetries of Belonging." The lectures were later published with the same title as the first in the series. I saw this project as another jewel in the crown of ISAP. This was also intended to attract an international audience to Zurich, like the annual Jungian Odyssey does every spring.

Spring Journal Books turned out to be very important for increasing ISAP's international visibility. Nancy Cater not only publishes the Zurich Lectures Series but also the Jungian Odyssey papers. I thought it could be important for ISAP that our analysts not only lecture and teach but also publish articles and books as much as possible in English. I was guest editor for an issue of Spring Journal titled *Symbolic Life 2009*, which is a collection of papers by ISAP analysts about symbols. This was in honor of Jung's famous lecture in London in 1939, "Symbolic Life," which was also used as the title of the eighteenth volume of his *Collected Works*. I think this alliance with Nancy Cater and Spring Journal is very beneficial for ISAP.

IM: Another connection has been with Steve Buser and his ability with electronic media...

MS: Yes, in November 2007 when I was lecturing in Chicago I met Dr. Steve Buser. He was training at the Chicago Jung Institute and asked me if I had ever considered doing video seminars. He thought perhaps training seminars broadcasted via the Internet could supplement training in out-of-the-way places. He lives in Asheville, North Carolina and was commuting to Chicago once a month for training in Jungian-oriented psychotherapy. I thought this was an intriguing idea for bringing ISAP analysts to a wide and far-flung international audience. So we began in 2008, and he set up a network from Asheville that would shortly broadcast to approximately 20 countries worldwide – to places like Iran, Japan, Russia, Eastern European Countries, South Africa, Israel, and so forth. When I travel nowadays many people mention to me that they watch these programs. This was also an attempt to make ISAP known. John Hill, Brigitte Egger, Andreas Schweizer, Paul Brutsche, Mario Jacoby, myself and other ISAP-analysts have presented lectures by video. The Asheville mailing list includes around 20,000 addresses from all over the world. The seminars are recorded and we have DVDs made, which are available for sale. To date, we've made over 30 DVDs. I take a few as gifts when I travel to other countries, and some of them have been translated with subtitles (into Russian, for example). The entire set forms a pretty solid course in Jungian psychology. That was an attempt to take the ship of ISAP further out on to the high seas of the international community. Many people have become aware of ISAP through the Asheville programs.

IM: 2009 seemed to be a busy year for you. I remember the publication of the *Red Book* in that year.

MS: You are right. The publication of the *Red Book* was an astonishing event for Jungians worldwide. We organized a public symposium at ISAP in late November. The house was full. Everyone was intensely interested to hear Ulrich Hoerni, Kathrin Asper, and Andreas Schweizer speak of their experiences with the weighty tome. Ulrich, who is Jung's grandson and the person most responsible for the publication of this important work, spoke of the history of the project. Kathrin and Andreas offered some preliminary estimations of the significance of the work for Jungian analysis today. In 2010, the Rietberg-Museum in Zurich organized an exhibition of the *Red Book* and other artistic works by Jung, in collaboration with the Stiftung der Werke C.G. Jungs (Foundation for C.G. Jung's Work). ISAP was invited along with the Jung Institute to offer guided tours by our analysts, and we also held an all-day symposium on the work and its significance. We worked together very well with the Jung Institute on this project.

IM: Coming back first to ISAP and the faculty: Did you change some things after taking over the presidency from Paul?

MS: One of my immediate goals was to cut down on the long meetings. With the officers, I proposed forming a steering committee, which would meet every week for an hour and a half and an ISAP-Council that would meet only once or twice per semester for the purpose of taking decisions. For longer discussions and strategic reflections, I proposed an all-day retreat once per year. I thought it was important to reduce the time spent in interminable meetings and discussions. People were losing patience and burnout was a topic on many lips.

The idea of finding a new location for ISAP was also in the air. I knew the Psychological Club was about to begin renovating their marvelous building on Gemeindestrasse, and my thought was to see if ISAP could find its new home there. The Jung Institute had originally been housed on the second floor of that building, and Jung's office was located on the first floor. So the building had a lot of history to recommend it, and I feel it has a powerful aura—this is, after all, where Jung himself gave his famous lectures and the rooms still are filled with the presence of historical figures that lectured there, like Emma Jung, Toni Wolff, Marie-Louise von Franz and so many others. When the Jung Institute heard about this plan, they objected. They saw it as favoritism toward ISAP on the part of the Psychological Club. In the end, the

Psychological Club's board decided not to offer the space to ISAP in order to maintain the appearance of neutrality. In the end, I think it is better we didn't move there. Emotionally and symbolically it's a good idea, but not practically. The space is actually too small for ISAP's needs. It's much better where we are now; there is more space for us.

One difficult issue to live with at ISAP during my time as president was the subtle and often not spoken but ongoing desire on a part of many participants to go back to Küsnacht. I felt like Moses hearing the Israelites murmuring in the wilderness about how much they missed Egypt. Oh, the fleshpots of Egypt, I thought—forget them! We are making our way to the Promised Land! This regressive wish blew at us like a headwind, preventing us from making progress in our necessary direction toward becoming fully actualized as an autonomous and independent society. I would say, "No, the Promised Land is that way, ahead; we have a new destination. Do not even think of returning to the old place of stagnation." It was never an open conflict, but it felt like a drag on our ability to think strategically. Today it is clear: we are not going back. There is no going back to where we came from. The only way is ahead. I am certainly not opposed to consider a union of sorts with the other group who remain behind in Küsnacht. But they need to come to us.

I can easily imagine a new configuration of people and programs: The Curatorium of the Jung Institute (a "Stiftung" [Foundation] under Swiss law) could foster and engage in research on the one side; a fully independent, democratically organized, IAAP member group made up of the combined faculty of the two present institutions could run the training programs. However, this is not possible as long as the people who caused the split in the first place remain on the faculty of the Jung Institute. Those people should be held responsible, as Heike Weis said at the "Emotions" Symposium (January 25, 2014)—it was the best comment of the day, in my opinion. As long as the perpetrators are not personally accountable for what they have done, the victims keep ruminating and wondering if it is their fault. Once they are tried and found guilty, people are freed from this frozen state. The Jung Institute keeps them on their faculty to this day. I don't understand why.

IM: This is still an ongoing issue with the Jung Institute. But you worked together on projects, as you said?

MS: Yes. In August 2008 Paul Brutsche and I suggested to Daniel Hell, director of the Psychiatric Clinic Burghölzli that we celebrate Freud's visit to Jung there in 1908. Daniel Hell liked the idea, and he said the Jung Institute should also participate as well as the Freudian institutes in Zurich. So we agreed to this plan, and he organized the day accordingly. The event went well. ISAP gave some lectures, the other institutes gave some lectures, and it was basically in celebration of Freud's visit to the Burghölzli. A spirit of unity was shown in an unexpectedly large turnout of analysts from the Freud Institute of Zurich, the Psychoanalytic Institute of Zurich, ISAPZURICH, and the C.G. Jung Institute.

IM: And this spirit of cooperation with others continued?

MS: Indeed. In 2011, together with Thomas Fischer and Ulrich Hoerni of the Stiftung der Werke C.G. Jungs (Foundation for C.G. Jung's Work), we organized a conference at the ETH of the topic "C.G. Jung im historischen Kontext der 30er Jahre: soziale, politische, und professionelle Aspekte." (These papers were published in the *Analytischen Psychologie*, vol. 168, Feb. 2012, and translated and published in *The Jung Journal* under the title "C.G. Jung in the Historical Context of the '30s: Social, Political and Professional Aspects.") Speakers were Geoffrey Cocks, Giovanni Sorge, Ann Lammers, Werner Disler and the historian, Georg Kreis. In the same year, ISAP marked the 50th anniversary of Jung's death together with the Jung Institute, the Psychological Club, the Schweizerische Stiftung Museum für Analytische Psychologie nach C.G. Jung (Swiss Foundation Museum for Analytical Psychology in the spirit of C.G. Jung), and the Stiftung der Werke von C.G. Jung (Foundation for the Work of C.G. Jung). Daniel Hell, Verena Kast and Allan Guggenbühl gave lectures—also the president (mayor) of Zurich, Corinne Mauch. That was a cooperative effort. On this level, as I mentioned before, with respect to the Red Book exhibition at the Rietberg Museum, the spirit of cooperation was excellent.

IM: Back to the ship and what is going on in the ship ...

MS: The basic mission is to run a first-class training program for Jungian psychoanalysts. This includes many lectures, seminars, colloquia and other learning experiences over two semesters each year. At ISAP, we have departments set up in twelve different areas. Each department has a head and is takes responsibility for offering courses and seminars as well as for keeping reading lists up to date. This is the basic program that is offered on the ISAP ship. It should also be said that we have many creative analysts who specialize in various areas of interest. The students receive the benefit of their expertise and enthusiasm for their topics of choice. There is also an experiential side of the training program. Weekend classes are offered in psychodrama, active imagination, dream and fairytale interpretation, body awareness, and so forth. Altogether this makes ISAP very special. Normally you don't find training institutes with so many specialties. So this is one of the features this makes ISAP extraordinary. The faculty is outstanding, and the students come from all over the world. And all of this takes place in the city of Zurich, the spiritual home of Jungian psychology. This combination lends ISAP an awesome attractive power, an aura of excellence.

IM: Thank you very much, Murray, for this informative and lively interview!

THE CLASSICAL DESIGN OF THE ISAP SHIP
Our Co-Presidency, 2012 to Date

MARCO DELLA CHIESA & ISABELLE MEIER

The ISAP ship was built under the presidency of Paul Brutsche, and then steered onto the high seas under Murray Stein. Our mission was to reinforce the ship and its classical form, to stabilize it and put it into the context of other vessels. We are still in the midst of this mission.

We took over the helm of a well-organized and functioning ISAP in March 2012. Collaboration with the Steering Committee (Stefan Boethius, Finance and Erhard Trittibach, Secretary) was characterized by mutual respect and willing support. We met with them every Wednesday at the Kunsthaus to discuss the tasks ahead. The local waiters got to know us and our wishes. Ursula Ulmer (Director of Admissions) often joined us, for which we were grateful.

In contrast to Murray Stein's presidency, and at the request of other ISAP Council (IC) members, such as Nathalie Baratoff (Director of Program) and Marianne Peier (Director of Studies), we increased the number of IC meetings per year and also extended the discussion time. Discussion was important for establishing a balance between the different views.

ISAP has faced great challenges that affected three areas. We moved with our ship in troubled waters, for three deep currents had seized it, and we needed the support of many colleagues and all our skills to calmly steer such a large ship. These three great currents were:

- Swiss federal regulation of the psychology professions ("PsyG")
- Cooperation with AGAP
- Interactions with the C.G. Jung Institute

We were united in the determination to offer an in-depth, comprehensive, and "Zurich-style" training in analytical psychology in Zurich. Ultimately the energy for the commitment and participation in the many different bodies arose and, in our view, continues to arise from this attitude. Without this largely voluntary commitment, ISAPZURICH could not continue to exist, as we have gratefully and repeatedly acknowledged. But we are also convinced that this commitment to ISAPZURICH brings a lot of satisfaction, joy, and enriching encounters and experiences along with the hard work.

The Federal Law of the Psychology Professions

In a democratically organized institution, information plays a central role. To establish the ship's approach towards the federally ruled domestic waters, we organized several meetings with our participants. This process began in 2012 for we knew that on April 1, 2013 a new Psychology Professions Act (Psychologieberufe-Gesetz) would come into force, which requires psychologists to complete psychotherapy training to be able to practice in conjunction with health insurance companies.

We heard from the C.G. Jung Institute that they wanted to split their programms into a path for "psychotherapist," and another for "Jungian psychoanalyst." The requirements for the former were lower than for the latter because the new PsyG criteria had greatly reduced them; the psychotherapists meeting these requirements were therefore already qualified to work with health insurers with this "lighter" training. The core of the controversy was sparked by the number of analysis training hours required under the PsyG (i.e., 150, only half of the 300 hours required by us).

We discussed this two-part model in the spring of 2012 with our participants, and it triggered some angry squalls. There was no way that depth psychology training could require only 150 hours of analysis; it is by no means enough and contrary to the Jungian concept of psychoanalysis. Clearly a Jungian could work psychotherapeutically, although always from an analytic stance, and would therefore need the full training. Others believed that the training could be split into two parts and that psychotherapists would, after qualifying, be sure to complete the rest of their analysis and work. We set up an e-forum to give everyone the opportunity to express his or her thoughts outside the meetings.

During the summer of 2012, a consensus emerged. Our "core business" is the training of Jungian analysts. If any of our candidates want to work in Switzerland as a psychotherapists qualified to treat under health insurance coverage, then they must complete our entire training. This view was presented at the extraordinary General Meeting in the fall of 2012 and approved by a clear majority. We took the position that we don't want to offer streamlined training; we want to include topics such as the "Psychology of Religion" and "Ethnology and Psychology," as well as require a thesis; we maintain that sufficient personal analysis is essential for a good professional practice—be it psychotherapy or analysis. In our opinion, the currently fashionable trends reduce therapy or analysis too much to the level of coaching, psychoeducation, or theoretical knowledge.

The storm abated, and we decided to pursue accreditation from the federal government with a classical program. With our emphasis on the standards of a classical training, we have set ourselves apart from most other Swiss psychotherapy institutions that will adopt the lower federal requirements.

Cooperation with AGAP

AGAP's membership consists primarily of analysts who graduated from the C.G. Jung Institute Zürich or ISAPZURICH. Most of them came from abroad and spent several years of their lives in Zurich. In 2006, the C.G. Jung Institute changed their international training to a block program, which required a candidate to be present in Zurich for only a few weeks. This led to heated discussions among AGAP members about training and the basis for AGAP membership. When the debate came to a head at the AGAP General Assembly in 2010 (Montreal), AGAP President Deborah Egger assured the members that the Executive Committee (ExCo) would seek an equitable solution.

How were we to move on?

One of many steps was a day-long workshop organized jointly by the AGAP ExCo and the CGJI Curatorium with seven delegates from each group. Gathering on October 6, 2011 at the Volkshaus, the fourteen delegates worked with the goal to develop the basis and the content for a cooperation agreement between AGAP and CGJI. They generated some suggestions, including the idea to form what would become the AGAP Membership Standards Board (MSB) to which AGAP, ISAP and the CGJI would delegate three representatives each.

Under Murray Stein's presidency, ISAP participants voted in favor of such a working group. Some months later the newly elected ISAP Council, under our leadership appointed Deborah Egger, Isabelle Meier and Erhard Trittibach to represent ISAP on the MSB. Chaired by AGAP Co-President Stacy Wirth, the MSB included also Kristina Schellinski, Sandy Schnekenburger (representing AGAP at large), and Daniel Baumann, Renate Daniel and Philip Kime (representing CGJI).

With hindsight, perhaps it can be said that we have little to show for all our efforts, at least in our opinion. The two day-long meetings on September 23, 2012 and January 19, 2013 resulted in a lot of discussion about AGAP membership requirements. The ISAP representatives appreciated CGJI's increase of block training in Zurich from two to three three-week semesters per year. We appreciated as well their proposal to require training analyses and supervision to take place with AGAP members whenever possible. However, finding that such changes would not be enough to uphold AGAP's distinguishing character and identity based on Zurich "immersion training," ISAP ultimately offered a bridge proposal that would involve some exchange of training resources.

Although the two MSB meetings produced bridge proposals from the other groups and convergence on some points, no agreement could be found on the crucial issues. So it was decided to suspend meeting for the time being and to hand the question of membership criteria back to the AGAP ExCo. The ExCo then proposed the AGAP Membership Convention, which tried to resolve the diverging interests of all parties represented in the MSB. In June 2013 the ExCo sent a draft to the leaders of ISAP and the CGJI with a request for feedback. While the ISAP IC agreed with the proposal, the ExCo received no reply from the CGJI Curatorium, so we were told. In July 2013, the AGAP ExCo announced the adoption of the Convention and made the document available online for the whole membership.[1] We regard it as a compromise between the requirements of the two institutions.

The years 2012 and 2013 showed that ISAP has a clear stance in its relationships with AGAP, the CGJI and the Swiss Federal Regulatory Commission. We remain true to our classical, Zurich-based training program. We are of the view that, in the midst of ongoing and rapid changes, a Jungian analysis of the postmodern individual can still be a valuable aid for orientation and self-development as well as for the relief and cure of mental suffering. The genuinely Jungian point of view with its emphasis on the symbolic and creative and with a focus on spiritual transformation that goes beyond symptom control, is our essence.

Interactions with the C.G. Jung Institute

In 2012 the IC began to independently develop ISAP's contact with the C.G. Jung Institute. The Institute had undergone a major reorganization in 2011, and mutual interest grew in whether we could come together or could enter into a relationship of some kind. The meetings were attended by Stefan Boëthius, Marco Della Chiesa, Erhard Trittibach and Isabelle Meier from ISAP and by Georg Elser, Marianne Meister, Renate Daniel and Annette Jörgens from the C.G. Jung Institute. The first meeting was held in the main station buffet in Zurich, a neutral place where the initial objective was to see if further contact was desired from both sides. Many of those present had not seen each other for years. Thereafter, subsequent informal meetings were arranged.

We were aware that, in addition to the opportunities that would arise from such an encounter, there were also risks and hazards. We steer our ships through the same waters; this was also clear to the representatives of the C.G. Jung Institute. However, we took a chance and agreed to a preliminary navigational course. Both institutes organized a large meeting of analysts on April 20, 2013 with the Austrian, Friedrich Glasl, a well-known moderator, at which various ideas and projects for closer cooperation were discussed. We raised the idea of a full merger with a new name, but this was met with little enthusiasm. Concrete ideas, such as a jointly organized memorial event on the anniversary of Jung's death on June 6, could be realized more easily. We decided upon an "Emotions Symposium" on past and currently felt emotions, and this was jointly organized by the delegates of ISAP and CGJI, under the leadership of SGAP, and with external moderators less than a year later on January 25, 2014. The processing of emotions proved to be much harder than expected at the outset; many analysts expressed emotions related to past events, such as anger, sadness, pain and disappointment. Analysts of the CGJI even brought new disappointments over AGAP's alleged new membership criteria for admission.

At the moment both institutions steer their ships separately through the waters. Both are active and dynamic; their ships look different because they have different identities. (Since January 1, 2014, ISAP dwells in a new, central location in an Art Nouveau style building.) Two active and independent ships can strengthen the position of analytical psychology in Switzerland just as well as they could with a common way forward.

NOTE

[1] AGAP Membership Convention. Criteria for Regular Membership. http://www.agap.info/Data/Files/agap_membership_convention.pdf (downloaded 14 April 2014)

OUR IMAGE IN THE WORLD

THE JUNGIAN ODYSSEY, 2006-2012

JOHN HILL

Beginnings

Beginnings can be difficult and the Jungian Odyssey (JO) was no exception. Shortly after ISAP emerged out of the long and painful conflict at the C.G. Jung Institute in 2004, the new community assumed there would be a Summer Intensive, similar to the one we had known previously. We all knew it would have to change into another structure and, if possible, set at another time. Toward the end of 2004 a committee was formed. Cedrus Monte was appointed chairperson and Denise Bloom, Sasa Boethius, Eileen Nemeth, Connie Steiner, and John Hill were responsible for booking the various resorts.

At one of the preliminary meetings, it occurred to me that we had suffered some kind of paralysis in the long struggle at the Jung Institute. We needed to move and so the word "odyssey" came to mind. Cedrus continued to search for an appropriate title and we soon decided on "The Jungian Odyssey." The proposed conference now had an identity with a mythological dimension that would inspire us and give us energy to undertake this difficult endeavor. Mindful of the Homeric model, we were relieved that the war of Troy was over; we would build a ship that would survive the winds and tides of fortune, hopeful to reach Ithaca. We were not sure what "Ithaca" represented but looking back over the years, ISAP has become known to the world through the Jungian Odyssey.

What kind of materials would be suitable for our ship and when and where would be its first destination? It soon became clear that we were not ready to set sail in 2005. After much debate the committee finally decided on summer 2006. Unfortunately this meant that our first conference would be at the same time as the Summer Intensive at the Jung Institute. Later I learned this was seen as a hostile act. I can honestly say this was not our intention. At that time both the Jung Institute and ISAP followed the common university semester schedule. If the Jungian Odyssey were to succeed in any way, it would have to be attached to the end of the summer semester, so as not to disrupt it.

Next we struggled with the venue. It was clear that the very notion of an odyssey implies no permanent residence. We would move from venue to venue. It would take place each year in different regions of Switzerland, affording immersion in landscapes that influenced Jung's sense of psyche. Our conference would also be a retreat, each one infused with the atmosphere of particular *genius loci*. Jung once observed that *genius loci* may be "elusive, yet nonetheless ... present as a sort of atmosphere that permeates everything."[1] Flueli-Ranft, a tiny village located in the heart of Switzerland, was our first choice. This was the home of Brother Klaus, patron saint of Switzerland. Klaus was a farmer, soldier, local judge, and spent the last twenty years of his life in a hermitage at the bottom of a steep gorge. C.G. Jung and Marie Louise von Franz saw him as a shaman of his time and wrote about his unorthodox visions. Considering all the upheavals that we had experienced at the old institute, the committee felt the time was right to reorient ourselves on a vertical axis. In Flueli-Ranft you can only go up or down, thus we decided on this venue.

Throughout 2005 plans and schedules were created, tested, and sometimes abandoned. Some ISAP members wanted part of the conference at the ISAP building. We started working on the details (hotels, menus, transport, room arrangements, and catering to special wishes of participants) soon to realize that we were organizing two different conferences. A great idea but an impossible task! We hit upon a compromise. The opening festivities would be held at ISAP and the next day we would travel to Flueli-Ranft. Cedrus Monte took on the daunting job of shuttling our guests between the various hotels and Hochstrasse 38, miraculously getting them on the right trams at the right time.

We achieved a further compromise in balancing the horizontal and vertical levels of interest. We programmed the conference to cover two overarching themes: "Jungian Psychology Today: Traditions and Innovations" and "The

Quest for Vision in a Troubled World: Exploring the Healing Dimensions of Religious Experience." Given the complexity of this undertaking, conflict was inevitable. The area of competence concerning the Odyssey Committee and the ISAP Council (IC) needed clarification. In the autumn of 2005 we reached consensus; Stacy Wirth from the IC joined the JO committee, Cedrus Monte was responsible for our guests, and John Hill became chairperson.

The committee finally decided on the academic program. There would be two morning lectures, a choice of seminars and experiential workshops in the afternoon, and two excursions to allow the participants to take in the enchantment of the Swiss landscapes. We also included morning meditation, dream gatherings, films, and a theatre performance of the Jung-White Letters in the various odysseys.

You can imagine the relief felt among members of the committee when the JO ship was finally launched with an opening party on the July 8, 2006. During the week we did encounter moments of chaos, and we sadly lost one of our keynote speakers, our esteemed colleague Ian Baker. Nevertheless the first Jungian Odyssey was a resounding success. The participants' feedback was overwhelmingly positive. Many descended to Brother Klaus's hermitage to partake in a world of silence, interrupted only by birdsong and the babbling of the nearby creek. In this way they could experience the two overarching themes of the conference, expressing the tension of living in two worlds, the visible and invisible, a world of discourse and a world of silence. ISAP had launched a ship as part of an endeavor to regain the esteem that Zürich once had.

Clearing the Decks for New Adventures

In organizing the Jungian Odyssey, we could never assume that the previous one would be similar to the next one. We kept an open door to the promptings of the soul. We listened to its gentle voice, giving us clues where to go, what theme to choose, whom to invite, what hotel was most suitable. Indeed we could give useful advice to many tourist agencies looking for guidance on hotels with soul. Following our first Odyssey, the JO committee was reorganized. Several members of our team were not ready to continue after the exhausting work of the first years. Isabelle Meier became chairperson, Stacy Wirth took over the responsibility for our guests, John Hill remained academic director, and Katy Remark was our new mate, ready for all kinds of tasks.

The new team decided that the Odyssey 2007 would be in Gersau, a small village on the shores of Lake Lucerne. I quote my earlier description of this particular Odyssey:

> Now less ambitious in our aims, we had decided on the theme, "Exploring the Other Side: The Reality of Soul in a world of Prescribed Meanings." This again, was inspired by the place – first by the enormous, glimmering lake and distant shores – and then by a certain mountain meadow lying opposite Gersau, namely, the famous Rütli meadow. According to legend, it was here that Wilhelm Tell defied the foreign oppressors – and here was signed the 1291 pact that united the first three cantons and laid the foundation for the future Swiss Confederation ... [Our] theme included a focus on prescriptive rules, revolt, hidden meanings, and the discovery of private myth – all urging those present to explore alternative ways of being.[2]

We knew that this dramatic landscape had a transforming effect on Friedrich Nietzsche, Richard Wagner, James Joyce, and C.G. Jung. Resonant with the surroundings, the contributions of our guests, James Hollis, Rafaella Colombo, Waltraut Körner, Rhoda Isaac, and Christopher Bamford as well as the many ISAP presenters, had a similar effect.

Our destination for 2008 was Beatenberg, supposedly the longest village in Europe, consisting of one main street, perched on a high bluff overlooking the Lake of Thun. The theme of the conference was "Intimacy: Venturing the Uncertainties of the Heart." Our first guest speaker, Ursula Wirtz, opened the conference with an inspiring talk, "The Yearning to be Known: Individuation and the Broken Wings of Eros." Using poetic and clinical material, she explored the light and dark aspects of Eros. Love involves the need to be seen and mirrored, but we also may encounter love's opposite: the lust for power and personal advantage over the well-being of another. Noirin Ni Riain, the well-known Irish vocalist was our second guest speaker. Noirin encouraged us to listen with the ear of the heart. Her music resounded in our souls and on several occasions she inspired us to sing with her tunes to themes of love and intimacy. Not to be forgotten was Paul Brutsche's presentation on Paul Klee's lifelong connection to this beautiful area of Switzerland. Not only did we have a view of the Jungfrau but also of the Niesen, famed through Klee's many paintings of this mountain. Many other wonderful contributions elaborated on the uncertainties of the heart from a clinical, practical, neurobiological, mythological, and cultural perspective.

Our third conference was also an occasion to celebrate an innovative addition to the Jungian Odyssey. Nancy Cater, editor-in-chief of Spring Journal Books, agreed to publish the conference papers and was welcomed as a new member of the JO committee. In a preface to the first Odyssey volume, Nancy describes Spring's own odyssey, birthing in New York, relocating to Zürich, Dallas, Connecticut, and residing presently in New Orleans. Spring knew the perils of an odyssey, having been nearly shipwrecked when Hurricane Katrina devastated New Orleans in August 2005. Nancy writes: "Luckily, Katrina did not leave Spring all washed up or beached, neither did we by any means abandon ship. Instead, through hard work and much serendipity we managed to remain afloat and on a even keel, sighting more than one rainbow on the horizon."[3]

Since then Nancy has enabled us to bring out a new volume each year on the Jungian Odyssey. Forthcoming this year (2014) will be volume VI, "Echoes of Silence: Listening to Soul, Self, and Other."

Our port-of-call for the fourth Odyssey was truly a place of overwhelming beauty, a valley surrounded by towering mountains peering down on wooded landscapes interspersed with crystal clear lakes. In the Engadin Valley, Segantini found inspiration for his world-renowned paintings and Nietzsche completed *Thus Spoke Zarathustra*. Our hotel, the famed Waldhaus, a castellated *fin de siècle* (end of the century) edifice, has been owned and run by the same family since its construction in 1908. The hotel director wittily remarked, "C.G. Jung was last here in 1926 and we have been wondering why he has not come back since." During our entire stay, the hotel staff kept open

C.G. and Emma Jung's room, no. 221, which has remained unchanged since their vacation in 1926.

The title for the 2009 Odyssey was: "Destruction and Creation: Facing the Ambiguities of Power." In the preface to the second Odyssey volume, James Hollis writes:

> The essays that follow explore the paradoxes of our power and powerlessness in this spinning universe. Power itself is neutral, tasked to address, perhaps resolve, the dilemmas of life. ... As a species we ill tolerate ambiguity, preferring clarity and control. But the anfractuosities of life seldom cooperate, and we are left to wrestle with the paradoxes of power, powerlessness, and destructivity.[4]

The rich diversity of presentations on this difficult subject was remarkable. All the speakers wrestled with both the dark and light aspects of power and powerlessness. In a paper full of poetry, Josephine Evetts-Secker invited us to hold the tensions of paradoxical energies, beginnings and endings, nothingness and creativity. Paul Bishop focused on destruction and creation in *Zarathustra*, inciting us to visit the Nietzsche's house just below our hotel. David Tacey dwelt on the power of the dark side of life and the need to relativize our conceptions of good and evil. There were many stimulating presentations on this enthralling theme, to name a few: Dariane Pictet elaborated on the dark aspect of the Indian goddess Kali, Andreas Schweizer on the ancient Egyptian visions of death and renewal, Bernard Sartorius on the tension between the king and his court jester, Mario Jacoby and Katy Remark on the clinical aspects of power, powerlessness, and dismemberment, and Kristina Schellinski on the re-anchoring of souls in today's global world of uprooted cultures and peoples.

The Jungian Odyssey 2010 returned to Gersau with a new theme: "Trust and Betrayal: Dawnings of Consciousness." Murray Stein opened the conference with an intriguing presentation on God's betrayal of Job and further elaborated on Victor White's experience of betrayal by Jung. Stein concluded that naïve trust in an "idealized goodness" will eventually activate its opposite, as witnessed in Christ's cry of abandonment on the cross. Diane Cousineau ended her moving account of trust and betrayal in the Job story with clinical material: Sometimes we have to betray what we most cherish, but we must not give up trust in the self, despite betrayal. Don Kalsched gave a brilliant clinical description of the betrayal of innocence, as witnessed in the splitting phenomenon of the vacated body of an abused child. In cases of early trauma, "Trust is betrayed so badly that the reality of the world cannot be taken in."[5] Love and deep relationship can create a healing space bridging virginal innocence and the sufferings of the human condition. Several other important papers elaborated on the entanglements between trust and betrayal in marriage (Deborah Egger and Christian Roesler), in the analytic relationship (Allan Guggenbühl), in the role of grandparents after the collapse of Hitler's Germany (Joanne Wieland), and in Jung's relationships with women (Judith Savage).

Our sixth Jungian Odyssey took place in Monte Verità, Mountain of Truth. The theme of the conference was "The Playful Psyche: Entering Chaos, Coincidence, Creation." We knew this was a daring title, connecting two very different worlds, artistic imagination and hardcore science. The venue itself, a hotel perched on a hill overlooking Lake Maggiore in the Italian speaking part of Switzerland, expressed this tension. It was owned by the ETH (Swiss Federal Institutes of Technology), also known locally as "Science City." In 1899 a group of European anarchists, having rejected the norms and inhibitions of bourgeois society, created a community on this hill that would explore alternative ideas and lifestyles. We also had the opportunity to hear talks at Eranos, the famed site on the shores of Lake Magiore. The diversity of ideas and improvisations was so great that it would be impossible to do justice to any one of them. I quote a summary of the Jungian Odyssey series editors, written shortly after the 2011 conference:

> Carrying forward the spirit of the place, our keynote speaker, physicist, F. David Peat, launched the Odyssey asking, "Can physics dance again?" The professor of chemistry, Reinhold Nesper, delved into the hard science of chaos, complexity, non-linearity, and emergence, bringing the concepts within the grasp of our broad audience. Lisa Sokolov, the singer and profes-

sor of experimental voicework, began her lecture with a mighty, improvised song.[6]

Our Jungian guests, Beverley Zabriskie and Joe Cambray, delved into the riddles and play of synchronicity. Our guests could also visit many experiential workshops, intended to keep the balance between play and imagination and disciplined scientific thinking, to name a few: Mythodrama (Alan Güggenbühl), Body-Soul Connections (Inge Missmahl), Psychodrama (Marco Della Chiesa), Dreamscapes (Eileen Nemeth), Moving with Chance (Stacy Wirth), Expressive Drawing and Painting (Isabelle Meier)

The venue of the seventh Jungian Odyssey was the beautifully renovated Paxmontana hotel in Flueli-Ranft, site of the first Odyssey. Our title: "Love: Traversing its Peaks and Valleys." Ann Ulanov opened the conference with a presentation on "The Particularity of Love." Love often begins with a wide openness to the ineffable, but loss, mourning, even violence are never far behind. We do not love in general but love a particular man or woman, a particular neighbor, a particular animal, smell, body. Our task is to create a meeting space between the All and the Small. James Hollis in "The Eden Project Redivivus" reminded us that we are blocked from returning to paradise, even if projections assume that is possible. Power replaces love when accountability is lacking. In a new Eden project we must learn to listen, take on responsibility, and relate to the other as other. Mark Patrick Hederman compared human life to that of a butterfly. We are born into three wombs, that of the larva, chrysalis, and butterfly; the first is dark and containing, in the second, we are subject to the wounds and bruises of life, in the third, we may explore those wounds as openings to the life of the spirit. Lucienne Marguerat delighted all in elaborating on the importance of kissing in the drawings of Aloise Corbaz. There were also many talks on the great tales of love: Eros and Psyche, Tristan and Isolde, and Dante's Divine Comedy. Several speakers dwelt on the cultural perspectives of love: love in paintings, literature and religion. We also heard about Alpine songs of love, and last but not least participants were invited to attend workshops on "Body Grounding" and on "Love as an inspiration to create your own poem, picture, or clay figure."

The Jungian Odyssey has been a tale of adventure. Like the great odysseys of old, it emerged from a path of destruction, war, and loss. Despite the inevitable struggles we were able to launch the ship and set sail in July 2006. So far we have visited or revisited eight different places, each one representing a different focus of consciousness. Looking back, these "islands of consciousness" follow an inner tension expressed in the title and experience of the conferences, such as tradition/innovation, soul spontaneity/prescribed meaning, trust/betrayal, destruction/creation, science/imagination, and the peaks and valleys of love. In this way each adventure has created a narrative around a fundamental core of Jungian psychology. Humans need more than ever before a space to live with the tensions that inevitably emerge from the strife between opposites. Perhaps one of the greatest achievements of these conferences was to hold this space with care, containment, and a warm hospitality that made our guests feel welcome.

In 2012 Isabelle Meier and John Hill felt it was time to hand over the oars and rudder to our cherished colleagues, Deborah Egger and Ursula Wirtz, who together with Stacy Wirth, Katy Remark and Nancy Cater (Spring Journal Books) will no doubt dexterously guide the odyssey ship to new adventures.

NOTES

[1] C.G. Jung, "The Compilations of American Psychology" (1930), in *The Collected Works of C.G. Jung,* vol. 10, *Civilization in Transition*, ed. Herbert Read, Michael Fordham, Gerhard Adler, trans. R.F.C. Hull, Bollingen Series XX, 2nd edition §972 (Princeton, NJ: Princeton University Press, 1970). Here Jung uses the term "*spiritus loci,*" elsewhere he refers to the same phenomenon using "genius loci."

[2] John Hill, "Epilogue," in *Intimacy, Venturing the Uncertainties of the Heart*, ed. Isabelle Meier, Stacy Wirth, John Hill, Jungian Odyssey Series, vol. I, (New Orleans: Spring Journal Books, 2009), p. 259.

[3] Nancy Cater, "Preface," in *Intimacy, Venturing the Uncertainties of the Heart*, ed. Isabelle Meier, Stacy Wirth, John Hill, Jungian Odyssey Series, vol. I, (New Orleans: Spring Journal Books, 2009),p. xvi.

[4] James Hollis, "Preface," in *Destruction and Creation, Facing the Ambiguities of Power*, ed. Isabelle Meier, Stacy Wirth, John Hill, Jungian Odyssey Series, vol. II, (New Orleans: Spring Journal Books, 2010), p. xvi.

[5] Donald Kalsched, "Innocence. Benign or Malignant? Part 1: Clinical Struggle with Trauma in Light of Saint-Exupéry's *The Little Prince*," in *Trust and Betrayal. Dawings of Consciousness*, ed. Stacy Wirth, Isabelle Meier, John Hill, Jungian Odyssey Series, vol. III, (New Orleans: Spring Journal Books, 2001), p. 45.

[6] Isabelle Meier, Stacy Wirth, John Hill, eds. "Introduction" in *The Playful Psyche, Entering Chaos, Coincidence, Creation*, Jungian Odyssey Series, vol. IV, (New Orleans: Spring Journal Books, 2012), p. 2.

THE JUNGIAN ODYSSEY, 2012 TO DATE
URSULA WIRTZ

Odysseus was curious and full of longing for Ithaca, and this archetype of the explorer and seeker also inspired the souls of the new Jungian Odyssey crew. We set off in 2012 toward unknown shores with the purpose of discovering something new beyond the well-known paths of war, destruction and loss to become acquainted with uncharted lands and unimagined spaces of consciousness. Homer's Odyssey, a dangerous voyage replete with surprises, seemed an apt metaphor for our striving to find new wisdom to guide us on the path toward individuation. So we have now launched our vessel with Odysseus, the "Faust of the seas,"[1] in our company.

Beginning in 2006 I have visited each Odyssey port-of-call as a lecturer and have come to appreciate the spirit, the atmosphere of warm hospitality, and the intimate sharing and learning together as something very precious. But 2012 was a special year, the year in which John Hill handed me a metallic white ship's lantern and spoke words that still ring in my ears: "With full confidence and great pleasure I hand over this ship's lantern to Ursula. Ursula will carry this light into the furthest corners of the world, exploring new themes, and spreading the Odyssey message to those who are looking for deeper answers to life's riddles." Deeply touched, I decided to rely on my long experience of sailing in the Mediterranean to keep our Odyssey ship clear of dangerous cliffs and currents and treacherous straits like those of Scylla and Charybdis.

Fortunately with Stacy Wirth, an experienced committee member since 2005, and now with Debbie Egger and Katy Remark, we are a reliable new crew! Nancy Cater of Spring Journal Books has bravely accompanied the Odyssey journey, joining us by Skype as a committee member and the publisher of the Jungian Odyssey series. In her advisory and editorial functions, she is very present in our team and highly appreciated.

The Jungian Odyssey 2013 got off to a good start with 84 participants from 24 countries—some as distant as Mexico, Venezuela, South Africa, New Zealand, Japan, Taiwan, and China. We dared to sail beyond familiar horizons into an open sea, responding to the soul's beckoning. A favorable, perhaps sacred wind blew us to the shores of great silence. We anchored in the rural lowlands of Canton Thurgau not far from Jung's birthplace, Kesswil, in the safe port of Kartause Ittingen, a former monastery founded in 1150, nested picturesquely amid its own forests, farmlands, vineyards and orchards—an ideal *temenos* for our big questions: "Whence, whither, why?" Beyond all tension of opposites we immersed ourselves there for a whole week, paradoxically sparing few words on the subject of the great silence, listening for "Echoes of Silence: Listening to Soul, Self, Other."

The poet David Whyte accompanied our departure into this realm. His own compositions and echoes of other poets evoked those spaces of consciousness and spiritual dimensions that our special guest speaker, Lionel Corbett, described in the context of clinical practice: "Approaching the Totality: Self, Silence, and Presence in Psychotherapy." Lionel was fully aware of the paradoxical situation that compelled us to speak for a whole week on silence and quoted Lao Tzu: "Those who know, don't talk; those who talk, don't know."

Lecturer Dariane Pictet was inspired by the *genius loci* to explore silence in the history of Christian mysticism. Bernhard Sartorius provocatively asked about the "silence of 'God.'" Waltraud Körner shared her passion for meditative pilgrimage and amplified her insights with the legend of the eternally wandering Jew, illustrating the meaning of embodied spirituality for our life's pilgrimage. Craig Stephenson acquainted us with John Cage's observations about silence and synchronicity in music. Urs Mehlin lectured on the art of "A Long, Lonesome, and Tedious Way of Coming to One's Self." Kristina Schellinski undertook a different form of the eternal quest in her contribution of the "replacement child," Jung being one of them. While Ann Chia Yi Li traced the echoes and mysteries of ancient Taoist paths, Shirley Ma introduced us into Eastern spiritual practices of Yoga and Qigong to help us experience the unity of body and soul beyond words. Peter Ammann, another bridge builder, showed film excerpts of African healers, inviting us to attune to the ancestors and archetypal roots of Jungian psychology in Africa. Brigitte Egger focused on our questionable psycho-ecological behavior. Lucienne Marguerat contributed her views on the scandalous collective conspiracy of silence that shamefully concealed the exploitation of Switzerland's *Verdingkinder*, the so-called "contract children." Allan Guggenbühl provided a challenging counterpoint with his dictum, "*silence is unbearable: make more noise.*"

Thus we approached the supreme mystery of silence in manifold ways, also through an inspiring discussion of the film *Broken Silence* facilitated by Diane Cousineau Brutsche and Paul Brutsche, and also Ingela Romare's documentary film (*Rowing for Tranquility in Times of Burnout*) and through active imagination with Maria Bernasconi and Ursula Lenz-Bücker, through creative work in clay and painting with Katharina Casanova und Ilsabe von Uslar, and through meditation and the composing of Haikus with Ursula Hohler. Verena Bollag played meditative piano solos and helped us deepen our ability to listen. The musician Heironymus Schädler played his magic flute in the exquisite Baroque cloister church, building a bridge for us between sacred and profane and heaven and earth.

Harkening to the voices of the soul, we also practiced seeing soul in our visits to Ittingen's lush garden of roses and the austere old monks' cells. We admired the historic old town of St. Gallen and the overwhelmingly beautiful Baroque Abbey Library, one of the largest and oldest libraries of the world and a designated UNESCO World Heritage Site. Enjoying a fondue picnic at the edge of the forest in the middle of the week, we were suddenly overwhelmed when a wild thunderstorm blew in and rewarded those who braved it with a full rainbow shimmering against a quiet, eerily lit evening sky.

Enriched and ensouled we have set sail again to explore new territories. With hoisted sails at this writing, we are underway to our next port, Grindelwald, also known as the Eiger or Glacier Village as it faces the infamous and world-famous Eiger North Face, and the majestic Jungfraujoch, which boasts Europe's highest railway station at 11,332 ft. Nestled in the security of the romantic Hotel Schweizerhof with its promising wellness oasis, we want to wrestle with *The Crucible of Failure,* hoping that we do not to fail in this endeavor because in nautical language "to fail" means the splintering of wood, the shattering of a ship against cliffs and rocks.

In Grindelwald we want to get to the root of a very painful topic that reveals something more about the *genius loci,* where courageous climbers risked their lives scaling formidable heights to successful ends. We also want to find out about others who tragically failed. Failure also belongs to our fundamental human condition as Augustinus has stated: *"fallor ergo sum* (I err, therefore I am.)"

Faced with global failures in political, economic, and ecological systems our culture nevertheless despises failure and perpetuates the myth of perfection, progress, and success. We are setting out to overcome the fear of failure and will circumambulate the theme with the intention to befriend failure as a given truth and gift, trusting that our failures may trigger insights and broaden our consciousness about ourselves and the world at large.

For the first time the Odyssey retreat will set aside time for a special *temenos*, a protected space for the sharing of personal experiences, insights and questions related to failure. Facilitated by an experienced analyst and contained in mutual respect and confidentiality, we hope that an open exchange on this life vicissitude can deepen our Odyssey spirit of community. The *temenos* offers an opportunity to explore the paradox of failure and its inevitability in life. Perhaps it will turn out to be a space for mourning, celebrating, and sharing our failures as stepping stones to wisdom, as signposts that reveal who we are and who we are meant to be.

Our dive into the transformative power of failure will begin with keynote speaker Polly Young Eisendrath, who is also an engaging mindfulness teacher and clinical associate professor of psychiatry. We will also be led by our special guest lecturer, Renos Papadopoulos, professor at the Centre for Psychoanalytic Studies at Essex and Director of the Centre for Trauma, Asylum, and Refugees

As for future ports, for 2015 we envision a place in the heart of Switzerland, Hotel Seeblick in Emmetten. Perched on a cliff above the captivating

Lake Lucerne and set against the backdrop of towering mountain peaks, it promises to be a perfect place to gain new perspectives on humanity: *On the Brink: Transitions and Turning Points*. Faithful to our call to reclaim a more integrative, holistic exploration of soul with its paradoxes and complexities, we have invited Iain McGilchrist, an eloquent philosophical writer and psychiatrist, well-known for his interdisciplinary scholarship. He will offer key insights into the binary division of our brains and how this laterality shapes our culture and consciousness. He will be joined by Eva Pattis Zoja, who is an expert at sandplay therapy in vulnerable communities, working with children in acute crisis situations.

All that is left for us to do now is to pray for favorable winds that allow this year's sailing to become an adventure through which we may "perceive whatever holds // the world together in its inmost folds." (Goethe, Faust I)

NOTE

[1] Ernst Bloch, *Das Prinzip Hoffnung*. Gesamtausgabe, Bd. 5 trans. Ursula Wirtz (Frankfurt am Main: Suhrkamp, 1959), S.1202.

THE ZURICH LECTURE SERIES (2009 TO DATE)

MURRAY STEIN

The Zurich Lecture Series was launched in 2009 as a joint venture between ISAPZurich and Spring Journal Books. John Hill inaugurated the series with lectures drawn from the book, which was published in 2010, titled *At Home in the World: Sounds and Symmetries of Belonging*. This was followed in 2010 by Paul Bishop's lectures, *Reading Goethe at Midlife: Ancient Wisdom, German Classicism, & Jung*; in 2011 by Josephine Evetts-Secker's lectures, *At Home in the Language of the Soul: Exploring Jungian Discourse and Psyche's Grammar of Transformation*; in 2012 by Ursula Wirtz's *Trauma and Beyond: The Mystery of Transformation*; in 2013 by Toshio Kawai's *Haruki Murakama and Japanese Medieval Stories: Between Pre-Modern and Postmodern Worlds*.

Further lectures have been planned for 2014 (Warren Colman), 2015 (Craig Stephenson), 2016 (Paul Brutsche), and 2017 (Eva Pattis).

The purpose of the Zurich Lecture Series was stated by the editors, Nancy Cater and Murray Stein, in the Editors' Foreword to the first volume of the series:

"The Zurich Lecture Series in Analytical Psychology was established in 2009 by the International School of Analytical Psychology Zurich (ISAP-ZURICH) and Spring Journal Books for the purpose of presenting, annually, a significant new work by a selected Jungian psychoanalyst or scholar who has previously offered innovative contributions to the field of analytical psychology by either:

- Bringing analytical psychology into meaningful dialogue with other scientific, artistic, and academic disciplines;

- Showing how analytical psychology can lead to a better understanding of contemporary global concerns relating to the environment, politics, religion; or
- Expanding the concepts of analytical psychology as they are applied clinically.

For the series, the selected lecturer delivers lectures over a two-day period in Zurich, based on a previously unpublished, book-length work, which is then published by Spring Journal Books."[1]

The ZLS team from ISAP is made up of Ann Li, Erhard Trittibach, Penelope Yungblut and Murray Stein. The team is in charge of making arrangements in Zurich for the lectures, which have taken place in the Zunfthaus zur Meisen on Friday evening and in the Lavaterhaus next to St. Peter's Church on Saturday. The first weekend of October has been set aside annually for the ZLS in ISAP's calendar. Nancy Cater and Murray Stein as editors of the ZLS Series published by Spring Journal Books select the authors.

ZLS-dinner at the Zunfthaus zur Meisen, 2013 (Photographer unknown)

---

[1] Nancy Cater and Murray Stein, "Editors Foreword," in John Hill, *At Home in the World: Sounds and Symmetries of Belonging*, (New Orleans: Spring Journal Books, 2010) p. vii.

Ursula Wirtz lecturing at Zunfthaus zur Meisen, 2012 (Photo by Murray Stein, 2012)

Toshio Kawai lecturing at the Zunfthaus zur Meisen, 2013 (Photo by Ann Chia-Yi Li, 2013)

Murray Stein with Toshio Kawai, ZLS 2013, at the Zunfthaus zur Meisen
(Photo by Ann Chia-Yi Li, 2013)

John Hill delivering a Zurich Lecture, 2011 (Photo by Murray Stein, 2011)

Dinner at Kronenhalle, Zurich, 2013 (Photo by Ann Chia-Yi Li, 2013)

ORIGIN AND HISTORY OF THE MARCH CONFERENCES "MÄRZTAGUNGEN"

LUCIENNE MARGUERAT & BILLE VON USLAR

Most ISAP students make considerable sacrifices to come and study Jungian psychology in Zurich where analytical psychology originated and thrives. They come from remote places, leave their homes and families, and often struggle to make ends meet during their stay here because Zurich is very expensive. Judging from the numerous countries represented among the ISAP candidates, the reputation of C.G. Jung's psychology and of ISAP seem quite well-established in the USA, Canada, Australia, Japan and in many other countries.

But what about the German speaking students and auditors? How can we reach them more effectively? Is the decreasing interest in the German-speaking regions a case of "a prophet having no honor in his own country?" The Arbeitsgruppe Deutsches Programm (AGDP: Task force for the German program) was launched at the end of 2008 as an attempt to find answers to these questions. It is a fact that depth psychology is facing difficult times, in spite of the increasing concern about methods of psychotherapy that are based solely on empirical and statistical methods. More holistic approaches, including qualitative and quantitative analyses, are gaining recognition. We are convinced that analytical psychology has much to offer therapeutically as well as culturally.

The hard core of AGDP, consisting initially of Paul Brutsche, Maria Bernasconi, Katharina Casanova, Lucienne Marguerat and Bille von Uslar, met for prolonged discussions in the kitchen of Katharina's and Bille's office, aided by bread, cheese and wine. At times Doris Lier and Ursula Ulmer would attend. Later Marianne Peier and, more recently, Irene Berkenbusch have joined the group.

We wanted the March conferences to transmit our enthusiasm to the public and to open a space for lively encounters, giving everyone an opportunity to see and reflect on significant issues through the lens of Jungian psychology.

The following topics resulted from our yearly brainstorming:

2010   Evil: The Dynamic between the Abyss and Creative Energy
2011   Time: Subjective and Objective Experiences of Time – Past and Present
2012   Typical Man? Typical Woman? Sexual Identity and Sexual Orientation
2013   Family: Destiny or Choice?
2014   The Dream: Language of the Soul

From 2010 to 2013 the March conferences took place on one day including coffee and lunch breaks. In 2014 we decided to take a leap and try a two-day event in the new ISAP premises on Stampfenbachstrasse. At this point our team included the diligent support of three ISAP students: Kathrin Schaeppi, Marianne Tinguely and Margareta Ehnberg. The meticulous planning, directed by Paul Brutsche, has guaranteed that each conference runs smoothly.

We look for speakers primarily among our colleagues partly for financial reasons but also among well-known personalities concerned with Jungian psychology. Various presentations, workshops, and round table discussions provide different approaches to the topic. The tasty culinary catering provided during the breaks adds to the general well-being. An aperitif was initiated in 2014.

After each conference we received good feedback concerning the smooth and creative teamwork and the lively atmosphere. This has been a powerful incentive to continue for another year. At any rate we have many good themes waiting to be launched at future conferences.

The following excerpts from the introduction texts to each conference give a good idea about our work:

Evil: The Dynamic between the Abyss and Creative Energy (2010):

Evil is part of everyday life. Crime reports and shocking photographs are thrown at us by the media every day. What is it that makes humans evil? The crucial question is whether we are the ones responsible for evil or whether non-human evil takes us into its power. Or are we at the mercy of our genes, our drives, or our environment? We assume we are free. But if we do have free will, why don't we choose to do good deeds? What is it that draws young people into street gangs? And, does evil possess a creative energy of its own?

Time: Subjective and Objective Experiences of Time – Past and Present (2011):

There is an objective, physical time that is more or less indisputable: We count hours, minutes, and seconds, we speak of morning, noon, evening, the coming night and the next day. And inevitably, we grow older. Subjectively, however, time affects our lives just as much. This is the time that speaks to our feelings. Time can fly or stand still, can be wasted, can seem interminable or way too short. We can play for time and try to turn back the clock. Time is a cosmic wheel or a trickle. Time heals all wounds and time is money .... What is it in us that keeps wanting to re-shape time; what is it about time that makes it so malleable?

Typical Man? Typical Woman? Sexual Identity and Sexual Orientation (2012):

The characteristics and differences between men and women are examined with regard to self-image, sexual identity and sexual attraction in light of C.G. Jung's psychology.

What is it that makes a man a man and a woman a woman? The spirit of the times has undeniably changed the way we look at the differences between men and women. Does it mean anything to be a typical man or a typical woman? This addresses the very foundation of sexual feelings. Besides the biology of sex, what is the role played by archetypal factors and cultural influences? Various issues such as homosexuality, bisexuality, and transsexuality will be discussed.

Family: Destiny or Choice (2013):

Everyone has a family, a mother, a father, ancestors, relatives… whether we know them or not. Our family, present or absent, plays a determining part in our bodily and psychic identity. Those who shape us, however, are not only our biological relatives and in-laws. We are part of various ideational and social systems that function in a family-like way.

To have parents, siblings and children belongs to our primordial human experiences; they are archetypal relationships. Even those who were raised without parents may encounter a mother or a father in the outer world as well as within themselves, another person, a group or in an idea, a dream, or nature….

The Dream: Language of The Soul (2014):

In our sleep, our dreams are real, whether they show us a harmless everyday scene or a shattering experience. For thousands of years humankind has searched for a meaning in the language of the soul. Beginning with Freud and Jung up to the present, one of the foremost issues has been to find ways of grasping the meaning of dreams and of using these skills within the therapeutic process.

Some dream motifs can be found everywhere in animals, buildings, landscapes, and primordial human situations. Dreams may sometimes mark a stage in life or anticipate profound transformations and transitions. They can clarify and heal. Dreams can be a source of life energy and inspiration. They can indicate hidden potentials, buried wishes, denied fears or family secrets.

Our team will continue to do its best to ensure that the March conferences show show the richness, diversity, and amazing relevance of the Jungian approach. And we hope to inspire the public and attract new students and auditors to ISAP.

Fig. 1: March Conference ambience (2014) at ISAPZURICH, Stampfenbachstr. 115, Zürich (Photo taken by Marianne Tinguely & Reto Casanova, 2014)

PERFORMANCE THEATER FROM ISAPZURICH

MURRAY STEIN

Performance art can bring the pages of literary texts to life. With this in mind, a small team of ISAP analysts (Paul Brutsche, John Hill, Heike Weis and Murray Stein) set out to create a theater piece from the correspondence that passed between C.G. Jung and Father Victor White from 1945 to 1960. The volume of *Jung-White Letters* (edited by Ann Lammers and Adrian Cunningham) was published in 2007 and contains the story of an important professional relationship and personal friendship between Prof. Jung and Father White (they signed their correspondences, "C.G." and "Victor"). Seven performances followed, and a video recording was made of the performance staged at the Widder Hotel in Zurich in 2009.

The DVD, which can be purchased from the Asheville Jung Center, also includes interviews with the actors who recount their experience in the roles. (Murray Stein speaks with Paul Brutsche "C.G. Jung," John Hill "Victor White," Heike Weis "Soror Mystica," and Ann Lammers "the Narrator".) All persons involved collaborated in the construction of the theater piece from beginning to end. Fellow ISAP analyst, Urs Mehlin, offered helpful tips on staging and on the performance efforts, which were incorporated and much appreciated.

Then we performed *The Red Book*! This famous work was published in 2009 and offered us the daunting prospect to imagine creating a stage performance of scenes from Jung's literary account of his inner journey and self-exploration during his midlife period. Never minding a challenge, in 2010 the intrepid ISAP Theater Troupe, consisting of Paul Brutsche, John Hill, Dariane Pictet, and Murray Stein, set out to select and dramatize seven key scenes from Jung's *Liber Novus*.

This would feature Paul Brutsche, who once again took the familiar role of C.G. Jung. John Hill divided himself theatrically among the roles of Elijah, Izdubar, and Philemon, Dariane Pictet incarnated Salome and Soul, and Murray Stein narrated (as "Hermes"), tying the scenes together into a coherent story. After a multitude of meetings and long rehearsals *Scenes from The Red Book* was brought to the stage – in Zurich (three times), St. Petersburg, Paris, Copenhagen, Qingdao, and Taipei. Audiences responded enthusiastically in all of these disparate cultural locations.

For the cast this was deeply gratifying. Diane Brutsche acted as the technical expert for the first performance, after which Lucienne Marguerat joined the team in this key role and took responsibility for managing the sound, lighting, and musical interludes for the performances. On two occasions, Erhard Trittibach assisted her. Their contributions were critical to whatever amount of success we garnered from our theatrical efforts. Diane Brutsche and Jan Stein were in charge of make-up, which gave depth to the cosmetic features of the actors, and supplied generous moral encouragement, which furnished them with heightened confidence in their roles.

It seems that the performances did bring the texts to life in a new and convincing way for many people.

I invited the actors from both theater pieces to comment freely on the following questions:

- How did performing these Jung-White letters and the scenes from The Red Book affect your understanding of the works and of the authors?
- Did learning these lines and performing them have an effect on your own life?
- What do you think has been the effect of these performances on audiences?
- Have you thought about an acting career?
- Should performance art be included as a part of the ISAP training experience?

Paul as C.G. Jung in *Scenes from The Red Book*

Paul Brutsche played C.G. Jung in *Scenes from The Red Book* and *The Jung White Letters*:

Twice I had the chance and the challenge to play the role of C.G. Jung. I imagine that I was chosen for this role because I can easily imitate Jung's pronounced Basler intonation when speaking English since I was born and raised in Basel myself. First of all, playing the role of Jung has brought me closer to Jung as a person. I discovered him as a passionate human being, personally involved in his ideas, fighting for them with heart and commitment in his discussions with Father White, sometimes even losing sight of his opponent and the present moment. I could feel how psychological thinking was not only a matter of detached academic reflection for him but also something that included values, convictions, and therapeutic commitment. In *The Red*

Book I met a person who was vulnerable, aware of his loss of soul, a very touching person—human and authentic. Jung, whom I had always appreciated for his ideas, his concepts and his approach to the psyche but who had been somehow difficult to reach and to like as a person, became a convincing and approachable human being to whom I could form a warm attachment. I could feel his inner turmoil, his struggles and his doubts, and his desperate longing for his soul buried under the former scientific ambition.

Thanks to playing some scenes from *The Red Book*, I also discovered Jung's amazing artistic skill as a writer and as a dramatist. Many dialogues with figures of this inner drama are so well developed and psychologically so accurate that one wonders: Where the hell did he get all that from! It gave me a glimpse into his multifaceted genius that revealed not only an outstanding scientist, therapist and thinker, but also a man with amazing artistic skill and poetic expression.

Learning these lines affected me so that I better understood the importance of a real and honest dialogue between emerging fantasies and myself. I learned something about active imagination. And I rediscovered something about creativity: that new personal insights originate from inner images that are received from an unknown source and expressed by a responsive ego.

I think that the effect of these performances on the audiences has been similar to the effect they had on us performers. I heard from many who had attended our performance that they had been moved. The performance had given them new access to Jung as a man of flesh and blood. They were touched by the personal drama between Jung and Victor White and its peaceful resolution, and they were impressed by the depth and richness of Jung's Red Book. Many seemed to be motivated to read The Red Book more carefully with the new perspective that there was, after all, something of value to be discovered and understood.

I have never thought about an acting career. But as I had been involved at the end of my school days in Basel at the age of 20 in a performance of Friedrich Dürrenmatt's *Ein Engel kommt nach Babylon* (*An Angel Comes to Babylon*), where I had the main role of the beggar Aki. I had at least gotten very close to such an idea. I have always liked acting and imitating people and animals, to the great displeasure of my wife.

To include performance as a part of the ISAP training experience is a wonderful idea, but, being aware of the amount of work a performance requires, I don't believe in the possibility of such a thing, unless the students would organize a performance themselves and unless it could be realized within the group itself, as John Hill offers with much success in his frequent fairy tale enactments.

****

John as Izdubar in *Scenes from The Red Book*

John Hill played Elijah, Izdubar, and Philemon in *Scenes from The Red Book* and Fr. Victor White in *The Jung-White Letters*:

Let me start with Question Four. I was very involved in school acting, but never seriously thought of becoming an actor. I have always been drawn to drama, probably because of its power to transform. At midlife I trained in psychodrama and since then have specialized in fairy tale drama. At sixty, I needed no encouragement to accept the role of Victor White and, later, those of Elijah, Izdubar and Philemon.

Having been raised in Catholic Ireland and having studied scholastic philosophy at university, I could easily identify with the role of Father White. In his encounter with "the mighty Jung" (letter from 16th of June, 1947, containing the Soror's vision), I could sympathize with his initial enthusiasm, respect, and affection for Jung and also with his later anger and despair in not being able to find common ground. Having enacted this dramatic dialogue, I saw that the differences between Jung and White were unbridgeable. This was a conflict that transcended father/son transference. White's academic formation was scholastic, Catholic, and communal; Jung's schooling was from the Enlightenment, Protestantism, and thus, individualistic. Both antagonists were locked into a cultural conflict that prevented reconciliation despite their affection for one another and their sincere intention to reach agreement. I found myself in sympathy with White's distinction between a psychological perception and a philosophical concept (letter from 5th of April, 1952), a distinction that Jung failed to grasp in their dispute concerning privatio boni.

Learning lines and acting roles furthers flexibility in thinking, permits indulgence in a variety of emotions, awakens the creative imagination, and heightens awareness of physical presence. In performing Victor White, I gained a freedom to perceive Jung through non-Jungian eyes, a healthy perspective for one deeply attached to the Jungian community. The dialogue between Jung and White can only be the beginning of an unfinished project that attempted to bridge differences between Catholicism and Protestantism, Aquinas and Kant, and scholastic metaphysics and empirical psychology.

Our theatrical performance of The Red Book involved a sophisticated combination of dialogue, action, costume, music, and images. Embodying the roles of Elijah, Izdubar, and Philemon was a challenging experience. The prophet, the warrior, and the magician are powerful archetypal figures. In enacting the roles, I had to invest a huge amount of energy to make them come alive. At times I felt schizoid, somewhat dissociated, and even inflated.

When I had finished playing each role, I needed some time to recover in order to click into the next one.

It was always helpful to engage with Jung on stage. Enacting the roles, I had to reprimand him when he resisted me, scold him when he avoided me, ridicule him when he got too smart, and sympathize with him when he suffered. Reflecting on the ordeals of these encounters, I could only admire Jung's resilience to endure the conflicts of his tortured soul, evidence of a differentiated, flexible, and mature ego, capable of surviving dissociation and facilitating transformation.

The performance of The Red Book has influenced me in several ways. I became intimately aware of the archetypal core of certain complexes. The dramatic dialogue between Jung and his antagonists initiated a process that helped tame, humanize, and redeem complexes from their schizoid shells. It was also a wonderful opportunity to participate in a communal project that has enriched collegial friendship in a community that otherwise encourages individual work.

Judging by the final applause, each performance must have had an overwhelming effect on the audience. Obviously a Jungian audience was more sympathetic to our theatrical presentations. A few non-Jungians attendees found some parts of The Red Book difficult to follow and could not appreciate the "liver scene." As Victor White, I felt great support from the audience. I am not sure about how our fans connected with Elijah, and I think they were more enthralled with Salome; they seemed to be fascinated with Izdubar, the warrior-god from the East; they laughed with Philemon the sophist, and they were in awe of Philemon "in the white robes of the prophet." In China, one woman refused to talk with me, as the mere John Hill, after the show. He was nothing compared to "the great Philemon."

I don't think a theatrical performance of Jungian material should be included as a partial fulfillment of the regular training program. Theatre is play and entertainment and should remain free from obligatory requirements. Nevertheless I would recommend that each training institute take up such projects. Theatre is a great way to experience the existential conflicts that Jung encountered throughout his long career. Doing psychodrama, I soon discovered that most candidates have an aptitude for Jungian psychology and possess a natural talent for acting, probably due to imagination and their deep connection with the soul. Drama breathes new life into Jungian material, celebrating Jungian psychology as an authentic and convincing alternative to the personal and trendy psychologies of our day.

****

Dariane Pictet played Soul and Salome in *Scenes from The Red Book* and the Narrator and Soror Mystica in *The Jung-White Letters*:

My experience with *The Jung-White Letters* as narrator (twice) was at first challenging, since I was stepping into someone else's shoes in a performance already set and rehearsed. My concern was thus more technical, less about performing a role and more about having adequate diction and voice projection. Yet, listening to Jung and White in rehearsal and before an audience enormously increased my understanding of the two men and the struggles they were facing on the level of human friendship as well as their intellectual struggle and jousting. With the role of Soror Mystica, which I performed only once with about three hours to prepare, I felt that the time pressure forced me to focus intensely on my part without much time for self-criticism and narcissistic anxiety with the result that it just "happened" in a flowing spontaneous way, without much reflection. And I was so intent on the part that I had no idea how I performed it. Much to my surprise and joy, I was told that that Jung's Soror Mystica had indeed appeared, ever so briefly. I also noticed that I was freer then I was when I was younger; I was less in the way of the performance.

It had the effect of reminding me how much I loved the theatre and how much I missed being able to have it be part of my life. I felt that my journey has had many meanderings, yet it has culminated in an uncanny, coming together-of-different-sides-of-myself in the most unexpected way.

Dariane as Soul in *Scenes from The Red Book*

Members of the audience shared with me the sense that the book was lifted off the pages and touched them in a three-dimensional way. It is so very different to read a text and then to hear it, let alone see it. We are moved in a very different way. Performances can get to us, bypassing defenses and preconceptions. Jung talked often about the reality of psyche. Soul can be imagined as a very ethereal thing, but, when she appears as a woman with her sensitivities and character, it makes it more real.

I had an acting career in my youth, after attending drama school. For me, it is a strange and unexpected loop of life to be on stage again. I feel that drama school is something that should be on every school curriculum. It's about spirit moving the body, channeling emotions, embodying text. I could wax lyrical about the benefits it has had on my introversion.

I trained in BodySoul work with Marion Woodman, and I am very aware of how the body carries our complexes. To learn to express ourselves experientially through art, movement, voice, mask work, etc., loosens us. It complements analytical work by bringing symbolism into the body. I also found John Hill's fairy tale enactment very powerful. So yes, there is a Jungian approach to drama that is very rich.

****

Heike Weis played the Soror Mystica in *The Jung-White Letters*:

The "understanding" of the authors, Jung and White, came for me mainly out of my "inner work" in my psychotherapy and training. Performing was a new and interesting process, and I felt very much the real meaning of the positive Anima function by growing into the role and repeating and acting the visions of Barbara Robb, referred to in the letters as Soror Mystica. Performing triggered aspects concerning my own individuation and a sense of these great men struggling with a theme like privatio boni, which has deep meaning in my choice to become a medical doctor. Reading the book and performing the role of the Soror brought their theories closer to me.

How has playing the Soror Mystica affected my own life? I found my husband! Beside this important fact, I learned more about being in a role. A persona to step in and out of, and to feel the growing presence of the Self as the source in the depth, which will always be the real motivator for bringing life's potential into action. This was helpful in finding more self-assurance and self-connection. Very important:

Performing the visions of Barbara Robb motivated me to increase my own active imagination! To become much more aware of my own inner process. The effect on the audiences? I witnessed very complex and variegated projections and reflections. Certainly the human depth of these two men touched the

Heike as Soror Mystica in *The Jung-White Letters*

people in the audience, and through the figure of the Soror they witnessed the dynamism of the psyche, including the effect and meaning of synchronicity. Many were really touched by the care and time Jung and White took for sharing their most intimate thoughts and reflections and also their limitations.

As I mentioned in the interview recorded for the DVD, I am happy to become more aware about playing the Anima woman in life, in contrast to living as my own being, a woman, human and imperfect. I don't want to go further in acting. I don't want a career, but a life with soul!

Should performance included in the training at ISAP? Rather than focusing on acting alone, it could be included but along with other possibilities of soulful expression like writing, dancing, painting, music, etc. I find it really important that students are guided to undertake a path to connect with their own psyches and not so much to be under the influence of others (Jung, White, outer Mana figures, famous people, etc.).

****

Ann Lammers played the Narrator in *The Jung-White Letters*:

Performing the letters brought certain parts of the text into focus for me in a very deep way. This was something of a surprise because I had lived so intimately with all those letters for almost a decade before we started rehearsing the play. I discovered I could really hear Jung and White (in the voices of Paul and John) speaking to each other. This was most true in certain passages where the emotional intensity of the actors was especially high.

As narrator, I was merely reading the lines I had written for myself, based on my research into the relationship of Jung and White and their historical circumstances. The point of my role in the play was to help audiences grasp the context of the material presented by the other actors. I felt I was holding the frame for a sacred action, whose meaning was known fully only to God or the Self. That is probably why I decided I could not leave Zurich and return home immediately in early June 2007 after getting word of my brother's death. We were engaged in holy work, a liturgy, and I had to see it through.

Our audiences appear to have been deeply moved by *The Jung-White Letters*. Beyond that, I cannot talk about the effect of the play on those who attended. But I do recall that three of the grandsons of C.G. Jung, who were in the opening night crowd in 2007, commented movingly about Paul Brutsche's "channeling" of their grandfather. And it went further than that. One grandson told me in private he was astonished because a piece of gold-colored synthetic fabric that I wrapped around the shoulders of "the sick Jung" was the exact color of a blanket his grandfather had owned.

Ann as the Narrator in *The Jung-White Letters*

In high school I did as much acting as I could find time for. Later, in my late 20s, I spent much time onstage as an opera singer. The Amato Opera Theater on the lower East Side of Manhattan was my venue where I sang Nanetta in "Falstaff," Papagena in "The Magic Flute," and Susanna in "Marriage of Figaro." Then my life took a turn that made any further attempts at a singing career impractical. (My older daughter, who lives in NYC, is a professional actor and playwright.)

Others can address the needs of the ISAP training program better. But I would venture that, for some students, performance art might be a deeply effective modality for learning.

Performance cast of *The Jung-White Letters*

****

Murray Stein played The Narrator (Hermes) in *Scenes from The Red Book* and acted as stage manager-director for *The Jung-White Letters*:

From the beginning of our conversations about staging performances of texts such as *The Jung-White Letters* and later *The Red Book*, I was fascinated by the enthusiasm our cast showed for these projects. I had learned about the powerful effects that performance of literary texts can generate in players and audiences from friends and colleagues in Chicago, notably from Prof. Leland Roloff, who held a chair in the Performance Arts Department at Northwestern University, and from Roland Rude, who had played the role of C.G. Jung in a performance of *The Freud-Jung Letters* in Chicago in the 1980s. When I heard Paul Brutsche's voice in full Basler accent for his account of "C.G." during the rehearsals for the performance of *The Jung-White Letters* and saw John Hill in clerical garb, I was convinced that we could offer a most credible rendition of the originals. And so it was. And we were blessed with gifted, beautiful and brilliant female actresses! The energy of the players and their abiding commitment to the performances never wavered for a moment. We all learned from one another and the work was thoroughly collaborative from start to finish.

Murray as Narrator in *Scenes from The Red Book*

For both performances, the selection of passages to enact from the texts was made in discussion with one another as a team, which extended over many hours of careful study and exchange. This in itself was a profound learning process for all of us. Then, during the enactments, I found that I came closer than ever before to the personalities of the protagonists and, of course, especially to the person of Jung himself. Going over the lines again and again and committing them to memory brings one into the mind of the character being portrayed. Each of the actors felt the roles to the core of their being. They became the characters they were performing. From the side, I listened and participated, occasionally directing and coaching, offering suggestions and, at the same time, entering intimately into the unfolding dramas.

This coming closer, especially to Jung, has without doubt affected my teaching and writing. So often while lecturing, a line from one of the performances springs to mind and I find myself enacting the various characters as though they possessed me. Even as I take walks in the woods around my home, a scene or a sentence will pop into my consciousness, and I will find myself in a scene from one of the performances. I do believe participating in such roles could be of use to students at ISAP and help them to become more familiar with texts and personalities. Performance not only brings the text to life, it brings the performer to life within the character being played. It is a profound learning experience.

One of the great pleasures of participating in these performances has been the time spent in community with the cast members. We have had great experiences together, both in familiar and in the most unlikely and faraway places – Russia, China, the UK, France, Canada, Switzerland, Portugal, and Denmark. In all these locations, we of course represented ISAP. In the end, we are left with many stories to enjoy with one another and to tell our friends. All the many hours spent in rehearsal come to the point on stage and on the road and behind the scenes; we were a company of friends who were so grateful to have this opportunity to work as a team in such a meaningful way. As we took our bows following each performance, we were more than content that all the effort had been worth it, and our hearts filled with gratitude.

ISAP'S OUTREACH TO EASTERN EUROPE AND BEYOND

Jungian Pioneers in Lithuania
KATHRIN ASPER

Lithuania was the first Eastern European country to make contact with Swiss Jungians. In 1992, I flew to Vilnius and was able to give my first multi-day seminar. That was a memorable beginning and a deep and enriching encounter with the land and people, especially with a group of interested students of Jungian psychology. Over the years the contact continued and resulted in further visits to Vilnius.

Lithuania now has twelve trained analysts, and the Lithuanian Society for Analytical Psychology has been a member group of IAAP since 2010; it plans to become an official training body of the IAAP in 2016.

A training program in analytical psychology, founded by Prof. Gudaite, who received her training as an analyst in Chicago and Zurich, has run since 1996. This program was based on the efforts of psychiatrist Gina Knabikiené who, since 1991, has strove vigorously with her group to attract guests from abroad who teach analytical psychology. Under Dr. Gudaite's tireless and competent direction, and that of her colleagues, analysts are trained in a five-year program. The courses are offered by guest presenters from Europe and America and now also by "home-grown" analysts.

Since the founding of ISAP, the following colleagues have regularly given courses in Vilnius: Kathrin Asper, Irene Berkenbusch, Elisabeth Hartung, John Hill, Mario Jacoby (†), Kristina Schellinski, Murray Stein, and Ursula Wirtz.

Gražina Gudaite visited ISAP in 2012 to give lectures and seminars. Analysts in training attend lectures and seminars at ISAP and undergo regular analysis and supervision with ISAP analysts.

The first European Congress for Analytical Psychology was held in 2009 in Vilnius. It was a wonderful convention, organized down to the last detail with trips, cultural events, celebrations, and a wealth of high-level lectures and seminars given by native presenters together with others from Eastern Europe and around the world. Gražina Gudaite, now a professor of psychology at the University of Vilnius, has succeeded over the years to establish a global network and forge friendships across national borders.

We Swiss are proud to see how this country, which had so bravely fought for its freedom in 1990, has built up such an excellent training program and is now networked worldwide.

Back in 1992, everything was very different. After the Soviet occupation our colleagues told us about the seizure of the television tower of Vyautas Landsbergis, one of the founding members of Sąjūdis and, later, the first president of Lithuania. Those were fearful, critical days, and over and over again they told the story with tears in their eyes; the world heard us! Over the years, the Soviet dominated city became a thriving, self-confident Lithuanian city. In the country and in the city, old Lithuanian crafts began to blossom again; people sang Lithuanian songs and remembered Lithuanian festivals. Building after building was renovated, and many, many churches shone with their former glory. Over the years that I have traveled to Lithuania, I have grown fond of the country and made many friends.

To learn analytical psychology in Zurich or to do so in Lithuania (or any other Eastern European country) are two different things. I always called the Lithuanians, "Jungian pioneers." This is because they mastered the Jungian way of thinking for themselves, but they had to adapt it to their circumstances. Thus, in supervision, I've noticed time and again the large number of clients that were traumatized or recounted traumatic experiences in their family background. Under the Soviets, for example, family members were tortured or sent to Siberia or both. This made our Lithuanian colleagues receptive to the trauma psychology that spread quickly from 1987 onwards, which steadily refined the treatment methods over the next few years. Interesting discussions arose among these bright and studious colleagues about how Jungian psychology and the psychology of trauma might converge.

Also, in personal analysis and supervision, especially at a time when there were no analytical psychologists in their country, the Lithuanians had to establish pioneering routes. With great material sacrifice they traveled abroad to Switzerland, Germany, England, or Denmark to obtain the required analysis and supervision hours. They came for a few days each time and completed their hours through intensive schedules. Now they have enough analysts in their own country to give sufficient analysis and supervision. Despite all this, some of them continue to go abroad in order to further their training in analytical psychology.

According to my experience and conviction, to convey analytical psychology abroad, where there is no such tradition, should never be a process of "colonization." Approaches and regulations should not simply be glossed over. Jungian thought has to deal with new traditions, the particular historical background, the possibilities, and circumstances—remembering material factors—to connect with the new country so to be fruitful and able to develop on other grounds. This has been extremely well done in Lithuania. The group has become competent, very active in its undertakings, cosmopolitan and traditional at the same time. They invariably receive colleagues from abroad with big hearts and deep hospitality; you will always be met at the airport with flowers.

I thank my Lithuanian friends for the many years of friendship and sharing, and, together with my ISAP colleagues, I wish them all the best for the future. May their path ahead be showered with blessings! All the best—Viso gerausio!

****

Renaissance for Jung in Poland
IRENE BERKENBUSCH-ERBE

Under communist rule there was—at least, officially—no psychoanalytic research or treatment in Poland. It was the same in the other Central and Eastern European countries under communist rule. This all changed in 1989.

Immediately afterwards, psychoanalysis (above all, Jungian analytical psychology), enjoyed a renaissance in these countries, especially in Poland. In due course, an IAAP Developing Group was formed in the late 1990s. Its first regular course in analytical psychology was launched by members of the United Kingdom IAAP Member Group.

The Polish Association for Jungian Analysts has developed and increased ever since. Their goal is the foundation of a C.G. Jung Institute. Its main field of work and treatment for years to come will be the traumatic experiences from the Second World War. These still play a dominant role not only in the analyses of individual clients but also in the collective unconscious of the Polish nation.

Since the foundation of the Developing Group there have been training programs in Jungian analytical psychology validated by the IAAP at regular intervals in Kraków, Warsaw and Wrocław. Members of the IAAP have conducted training seminars, individual and group supervision, and experiential seminars.

Since December 2009, there have been two Router Groups[1]: the Polish Association for Jungian Analysis (PAJA), which is active in Kraków and Warsaw, and the Polish Association of Analytical Psychology (PTPJ), which is also in Warsaw. Having a Router Group in their own country is vital for Polish training candidates because for most of them it would be financially difficult to study in Switzerland or other European countries. As an IAAP member, I have worked with the PAJA Routers, mainly in Kraków but also in Warsaw, since the end of 2009.

Contact was first established by Gert Sauer, who has been active in the training of Routers in Eastern and Central Europe and Russia for years. Since my husband and I, out of personal interest, have enjoyed regularly attending Polish language courses in Kraków for years, it should come as no surprise that I became active in Poland. Until January 2013 I ran group and individual weekend supervision there in English and Polish, including work with sandplay therapy. The Routers have a sandplay therapy section within their group. The case reports there could always be delivered in Polish and that guaranteed higher authenticity. Three of the participating Routers, Gregor Glodek, Tomek J. Jasinski, and Malgorzata Kalinowska were awarded certificates in analytical psychology at the Copenhagen IAAP Congress in August 2013. Together with Krzysztof Rutkowski, there are now four certified Jungian analysts in the Polish Association.

I love working with Polish candidates because they are so highly motivated. Two Polish colleagues, Mr. Tomek Jasiński and Ms. Małgorzata Kalinowska, have initiated a lively cooperation for future projects. In February 2012 they founded a new training group, called Basic Training in Analytical Psychology, to which I have actively contributed with theoretical seminars since January 2014. It is a wonderful group. Its training purpose is to impart a basic knowledge of analytical psychology so that the participants can become PAJA candidates and consequently IAAP Routers. This enables them to complete their training as Jungian analysts.

I am very fortunate to have the ongoing possibility to be active in Poland not only because of the enriching contacts with Polish analysts but also because it gives me the chance to live my love for Poland and its people.

For the last two years, I have also been in touch with Professor Gražina Gudaitė in Vilnius (Lithuania). I have regularly run seminars there as well as group and individual supervision. My contacts have also been very cordial and enriching in every respect. I am very glad about the possibility of continuing this work in Lithuania.

****

Jung in Georgia[1]

JOHN HILL

Welcome to the Land of Prometheus

Georgia is a very beautiful country, a land between the East and the West, situated between two inland seas, the Caspian and the Black that border Russia, Chechnya, Azerbaijan, Armenia, and Turkey. Its wild majestic mountains, the Caucasus, reach heights above 5000 meters. Its impressive churches, monasteries, forts, and other edifices testify to a rich history, which includes a long struggle for independence. Originally it was known as Kolchis, the land of the Golden Fleece, home of the ill-fated Medea. Prometheus, who dared defy the gods, was bound to a Caucasian rock. The Georgians are a defiant people. They suffered cruel times under Roman, Byzantine, Arab, Mongol, Persian, Turk, and Russian subjugation. Georgia was one of the first nations to embrace Christianity and has preserved its own language with its unique script throughout the centuries, notwithstanding the many efforts to destroy its many cultural treasures.

Present-day Georgia is a proud, strong, and deeply spiritual nation. Its people are optimistic about the future, despite political and economic pressures. Since the war of August 2008, the economic situation has worsened. The shock of that war remains. The Georgians still face an uncertain future with Russian armed forces only 40 miles from Tbilisi. They remain a defiant people.

Visiting analysts cannot fail to notice the generosity and hospitality of the Georgian people, despite limited financial resources. The food consists of a lot of pastries of meat, vegetables, and cheese. Georgia is famed for its wine. On my first visit I was given a choice of homemade white or black wine. I tried a glass of black wine and it took me a day to recover. One specialty is a phallic-like purple object made of nuts and dried grape juice, a kind of energy stick, which the warriors ate while riding horses. It is a hell of a chew.

The Training Projects

In writing an appreciation of the future of Jungian psychology in Georgia, two words come to mind: endurance and tolerance. Due to the unsparing efforts of Professor Rezo Korinteli, a study group in analytical psychology started in 1993. This led eventually to the official founding of the Georgian Association of Analytical Psychology (GAAP) in 2002 by twenty Georgians with the aim to further Jungian psychology in their country and the hope that one day there would be a good number of Georgian analysts.

Cloister in Kasbegi, Georgia (Photo by John Hill, 2011)

The Georgian training has been a project shared mostly by visiting analysts from Paris and Zürich. Françoise Caillet was the first IAAP liaison who oversaw the program, and Zürich-trained Marina Conti became the first personal analyst. In March 2008 I was appointed IAAP Liaison, succeeding Francoise Caillet, who retired that same year. Finally in 2009 five GAAP members became IAAP Routers. Fortunately, enough funds were raised and together with an annual grant from the IAAP, the five Routers have benefited from intensive training. A program was started, consisting of monthly shuttle analysis with Zürich-trained analyst Jacques Mermod, who is fluent in Russian, and regular group supervision given by François Martin Vallas. Analysis and supervision were also available to the other GAAP members some of whom wish to become Routers.

In 2011 Marina Conti retired. I was faced with the difficult task of finding a Russian-speaking analyst who was not Russian. With the help of my colleague Gražina Gudaite we finally discovered Elona Ilgiuviene from Lithuania who, as a second personal analyst, now comes regularly to Tbilisi. In 2012 ISAP analysts Paul Brutsche and Erhard Trittibach gave the IAAP intermediate exams to five Routers and screening interviews to four candidates. In 2013 Leslie de Galbert from Paris joined our team as an individual supervisor. In the meantime we have now ten Router candidates and are well on the way to establish a Jungian outpost in this part of the world.

There is another important Jungian project in Georgia that has been initiated by our ISAP colleague Allan Guggenbühl. In 2007 Allan started the Nergi project, which is designed to train Georgian psychologists in group processes and mythodrama. The purpose of Nergi is to provide an adequate and protected framework for Georgia's many traumatized children from uprooted and devastated families who are victims of the 1991 and 2008 wars. The completion of this project was celebrated in 2013 and is now firmly established in Georgia and run by Georgian colleagues, several of whom are IAAP training candidates.

Why this Project?

Since the small beginnings in 1993, it has taken more than twenty years to arrive at this point on the journey. The IAAP training program provides personal analysis, clinical supervision, theoretical knowledge, and seminars on the development of an ethical attitude in the practice of Jungian psychoanalysis. Through this rich program and the dynamics that are activated among group members, the Georgians have an opportunity to refine their understanding of innovative approaches to psychology. Each member of the group can only do this in his or her own way. This entails respect for the Other, without loss of a sense of Self. The long process is a test of endurance, but the history of Georgia assures us that this virtue is not lacking in the Georgian character. If Jungian psychology is to have a future in Georgia, then it must build on those cultural foundations that have helped the Georgian people survive the onslaught of alien ideologies.

I have witnessed the same open and tolerant attitude among the GAAP members. I had the impression that they first welcomed Jungian psychology as a promising alternative to the communist ideology of their youth and young adulthood. The scars of that ideology, however, remain. Recently (i.e., February 2013), I was shown a graveyard that is a kind of pantheon dedicated to the poets, writers, and artists of Georgia. Most of them were shot by order of Stalin, who seemed to have been particularly brutal to the intelligentsia of his native land. One colleague remarked: "If you read more than three books in those times you risked being shot." In the beginning members of our group tended to apply the Jungian concepts in ways such that Jung himself might have said: "Thank God I am Jung and not a Jungian." Were they still afraid of saying the wrong thing? The group members who attended several IAAP congresses have now been exposed to a plurality of approaches in the field of contemporary analytical psychology. They now seem to be moving in the direction of a more open and free discussion, perhaps entailing inevitable conflict. I am confident that they can tolerate dissension without loss of unity and the overall purpose of their society.

Having had experience in several Eastern European countries, I am convinced of the importance of Jungian psychology in Georgia. Georgia and other Eastern countries are presently invaded by Western lifestyles, brands and ideas. If we are not active, many psychologists will have no option but to train in non-Jungian programs. Freudian, Behaviorism, and other schools are already active in Georgia. Jungian psychology upholds the reality of soul. The IAAP training will give our Georgian colleagues an in-depth approach to patients suffering from trauma, abuse, and lack of spiritual orientation. Jungian psychology can link the psyche of present-day Georgians with their older spiritual and cultural heritage. It is important that that heritage, forbidden under Soviet rule, be respected and reappraised within the context of a modern man and woman. Jungian psychology is the one most suited to further this task.

****

Every person counts and *Laima* lights the way
KRISTINA SCHELLINSKI

Professor Gražina Gudaite, who leads the Association for Analytical Psychology in Lithuania, invited me to Vilnius in June 2011 for a seminar on the trans-generational transmission of trauma, together with sessions of group supervision. Many of those training there in Jungian psychology have been working for many years as clinical psychologists and are knowledgeable about different approaches.

I was offering a seminar on trans-generational aspects of dream analysis, focusing on cases where "skeletons in the closet," in a very grim and real sense, appear in dreams and speak of events that happened often long ago. The seminar was titled: "When what can't be said shows up in dreams." This is an aspect of our analytical work where we bring to consciousness those dream contents that speak of trauma and complexes passed on over generations. Often this work entails addressing very difficult issues; it means working with the dream aspects of both victims *and* perpetrators, and their descendants, intra-psychically as well as inter-relationally.

My own roots take me back to Lithuania; my mother fled Klaipeda in October 1944, crossing the bridge over the Nemunas (then called Memel). When I was little, she recounted how the bridge had collapsed behind her—as had the terror of the Nazi regime in those final months of the Second World War. To visit Lithuania in 2011, which had once again become an independent state in 1990 after being occupied by oppressive forces for so much of its history, was both an enlightening and a solemn experience for me. I had not shared the experience of five decades of Soviet occupation, of the reprisals after the war, and the deportations to Siberia. Nonetheless, I could literally *feel* the accumulated trauma of the effect of two totalitarian regimes on individuals and society when I was walking the streets of the capital, looking into the eyes of people, young and old, passing by buildings where so much horror had been endured, so much blood lost in the last century. Images of destruction entered my dreams at night, but by day, I could see the vibrant creativity of a new generation dedicated to building, rebuilding, creating, and reconstructing.

The future was so young—it had just begun. The courage that is inherent in beginnings was still palpable.

"Every person counts" was the phrase that came to me when I met with colleagues-in-training in Lithuania, and also with those who had come from Latvia and Estonia. *Everyone* who was training to be an analyst in these countries that suffered so much collective and individual trauma, and *every person* they were seeing in analysis was trying to understand the accumulated trauma and the twisted paths of its transmission in an effort to transcend the suffering.

The seminar, comprising some 60 participants, was held in a room flooded with light in the Vilnius Archdiocesan Family Center right next to the church of St. Theresa and just a few steps from the Gate of Dawn in Vilnius. The image of an ancient wooden angel sculpture helped us focus our attention on the work ahead.

What I wanted to share, namely that Lithuanians were not only victims of the Second World War and Soviet occupation after the war but that many in Lithuania had also been perpetrators and that hundreds of thousands of Jews had been killed in Lithuania, was difficult for all of us, me included. Indeed, in many souls today, handed-down self-identities of victim *and* perpetrator were still active in the unconscious and sometimes even "at war" three generations later. Much care was needed when we attempted to acknowledge that the trauma of both victims and perpetrators had been handed on, unwillingly, unknowingly, unconsciously from generation to generation. It often became the task of the third or fourth generation to speak of what had been unspeakable before, to unearth unavowed secrets and to do so, if possible, before the painful psychic contents became unthinkable another one or two generations hence. Otherwise, the repressed or dissociated content may become impossible to even dream or imagine, eventually becoming almost unrecognizable and therefore ever more difficult to understand and treat. "When what can't be said shows up in dreams" was translated sentence by sentence into Lithuanian: I learned while listening to the sharing ... in a language that was once my mother's tongue but no longer mine – I felt the "nothingness of in-between."

From the cases that were being presented, it was evident that the first, second and third generations were suffering; often it was only a member of the third generation who was actually beginning to look at what had remained a secret. In the face of too much suffering, children had often been unable to talk to their parents about what had traumatized them. Parents had tried to protect their children from *knowing* so to keep them from danger, yet their suffering, their secrets, and their trauma were weighing down the souls of successive generations. Many of the clients of my Lithuanian, Latvian and Estonian colleagues were affected in the *third* generation by what earlier generations could not face, the trauma they had been unable to work through—so many of them faced in their souls, unaided, the psychic and physical consequences of rape, torture, deportation or killings, severe deprivation or abandonment. It takes an enormous effort to begin the "inner" work required to come to terms with what happened long ago, to become conscious of that with a view to prevent such horrors from repeating.

My message was also a hopeful one: that we could acknowledge the transmitted trauma, look at it in small doses in the group, in the analytical encounter in an effort to remember with an attitude that is open toward reconciliation, to attempt to transcend the suffering of generations of Lithuanians.

Half of my maternal grand-parental heritage from minor Lithuania, annexed in 1938 by Nazi Germany, and half from mainland Lithuania has made me sensitive to seeing that the only way to bridge a schism is to try and reach out and reconcile at the level of the soul and at the personal level whenever we are meeting the "other." Blaming the other, maintaining a victim attitude and demanding admission from the perpetrator that a wrong was wrought for the debilitating destruction, is no condition for healing.

What is needed is a soulful way of bringing light into darkness. An image of this came to me in a dream: I dreamt of a duck that carried a lit candle on her back. The duck is a symbol of *Laima*, the Lithuanian goddess of life and death, of birth and destiny, who gives and takes back everything....

I was impressed by the group of analysts training in Jungian psychology, by their clear diagnostic eyes and ears and their "seventh sense" or intuition; many could call on knowledge received while training in approaches other than Jungian. They also shared in a spirit of cooperation, for the roots of their analytical understanding reach deep into the wellsprings of the land to find its ancient symbols that somehow miraculously survived centuries of attempted annihilation. In the group supervision, they could feel the different contours and levels of a person's suffering presented as a "case," and they could share their own soul's vibrating resonances and echoes in their body. Psyche *and* soma speak also in the countertransference of what may not be communicated in other ways.

May *Laima* help every person in Lithuania who is helping another, who is carrying suffering expressed in soma or psyche towards the light of consciousness, be it their own suffering or the often unspeakable, unavowed suffering of parents, grandparents and great-grandparents. Every person counts and—with *Laima*'s help—can lighten the load.

****

Czech Republic, Lithuania and Estonia
URSULA WIRTZ

My commitment to train in Eastern European countries has to do with an emphasis on trauma in my work that started many years ago with the group in Moscow. My personal contact with those now graduated Jungian analysts is still lively and has deepened since the Russian translations of my books.

In 2004, the President of the Czech Society for Analytical Psychology invited me to Brno for the first time to give a lecture, seminar, and supervision.

The Czech group was originally founded in 1997 as a "study group," which later became a "Developing Group," and is now officially accredited as a study program by the Czech health system.

Our late ISAP colleague, Mario Jacoby, was there from the beginning and a regular guest lecturer and supervisor until his death. His loss has been a particularly painful experience for our Czech colleagues as he had prepared the ground for a deeper understanding of the theory of archetypes, the landscape of complexes, and the process of individuation. But what they have learned from Mario about the processes of transference and countertransference lives on very actively in their therapeutic practices. (There are now nine individual members of the IAAP and 10 routers.)

Since my first visit I was also deeply connected with the group, and in addition to giving lectures and seminars I have worked analytically with some members. I am impressed by the commitment of these analysands to engage in a long analytical process, which is a huge outlay. This includes, for example, travel by night bus from Prague to Zurich to work very hard with me through the next day, and then take the night bus back to Prague.

In supervision, trauma-specific questions were often prevalent—my book *Soul Murder* was translated into Czech in 2005—so for three years I taught the theory and practice of trauma therapy and the central importance of the imagination for the processing of trauma in my lectures and seminars.

Later, the group wanted to learn more about the interface between analytical psychology and spirituality, and we worked with great openness on questions about the importance of the spiritual dimension—subjects that were banned during the communist dictatorship in the underground.

Through my teaching in Lithuania, Estonia and the Czech Republic, I have gained deep insights into the patterns of structural violence in complex practical, political, and social problems of emancipatory identity development that require a lot of cultural sensitivity and readiness for critical self-reflection of one's own political and cultural context.

A great openness to other political-economic realities and necessities, an exploration of human rights and loss of meaning and value, and a willingness to critically question the cross-cultural transferability of analytical psychology and the so-called gold standard of our training are important for working with Developing Groups.

Both in supervision and analysis with traumatized people of Eastern European countries and also in my work as a trustee of an internationally networked war trauma foundation, it has become clear to me that there is no universal conception of what constitutes a trauma or how it can be treated. Symptom presentation and healing rituals are also predominantly culture-specific and dependent on the contemporary social context.

Among my ISAP colleagues, John Hill familiarized the Czech group with the relationship between psychodrama and depth psychology in 2006, and Murray Stein was in Prague for the first time in 2010 as the *Journal of Analytical Psychology* held a conference there.

For the Czech Republic, despite being known for its atheism, 2014 is a special year because the philosopher and theologian Tomas Halik won the prestigious Templeton Prize. (The other winners were Aleksandr Solzhenitsyn, Desmond Tutu, and the Dalai Lama.) Halik called for the "cultivation of spirituality" as essential to society, thus promoting an attitude that is also inherent in Jungian thinking. Murray Stein and I look forward to working again in Prague with the Developing Group in 2014.

****

Jung in Islamic Cultures
BERNHARD SARTORIUS

I have had the privilege to work during the past few years as a "Jungian" teacher in Tunis (Tunisia) and Damascus (Syria). I was an instructor in local "Developing Groups," and taught in the context of a scientific colloquium on the theme of "Depth psychology and Ibn al Arabi." These experiences have made me realize three things, which I would like to introduce briefly as yet-to-be-processed hypotheses:

Practical scientific exchange with colleagues and training candidates in an altogether unfamiliar cultural setting is essential in our time because it corresponds to the global psychic constellation, which is reflected in economic globalization. It is also indispensable because it shows us concretely that our western way to observe and describe the psyche is only one among many other equally adequate perspectives.

This experience—and I stress that it is experience, not just academic knowledge—confirms the insight that all meta-psychological systems of thought—the Jungian included—are essentially projections. In other words, to a large extent the unconscious needs of a given culture and epoch correspond, and thus none of them can claim, as we like to do as "Westerners," to capture *the* essence of the psyche and its life better than the others. Each culture developed a psychology that needs to express certain needs and repel others.

With regard to the specific encounter between Jungian psychology and the Islamic view of the psyche, I could see many points of contact, but also fundamental differences. "Jungians" feel at home in the common assumption with their Muslim colleagues that there is a trans-material soul and that psychology is concerned with what lies beyond ordinary consciousness and its level of insight. On the other hand—and there is a big difference in the experience for Muslims—the soul is, without a doubt, a vestige of God in man. The clarity with which this is assumed in Islamic cultures causes a continuously experienced de-centering of the individual's position in the cosmos. Psychologically, this manifests as a basic relativity of the importance of the ego and its needs, a relativization that is not verbalized—such as in Buddhism—but which is all the more noticeable.

These issues, which I, as a Western analyst, have experienced first-hand in Islamic regions, have caused a profound upheaval in my thinking and in my practice which, as I have learned from my colleagues there, is also the case from the Muslim view. We maintain a process of mutual questioning and enrichment.

NOTES

[1] Part of this contribution first appeared in German: John Hill (2011). "Einige Überlegungen zur Zukunft der Analytischen Psychologie in Georgien," in *Analytische Psychologie,* 165, (March 2011): p. 354-356. It was partly printed in English in the IAAP Newsletter 2013.

WHAT ROLE DOES ISAPZURICH PLAY IN THE PUBLIC SPHERE?

ALLAN GUGGENBÜHL

The youth seems stubborn. He avoids my gaze and sits tensely at the table. He feels unwell in his own skin and now, on top of it all, by order of the juvenile court he has to go to this psychologist. They want him to reflect about his delinquency. He is held to be a hopeless case. I ask him about his mood: no answer. I want to know where he comes from: no answer. Finally we sit in silence and he, bored, begins to look around my room. How can I reach this boy? Suddenly his face brightens; he is amazed and he opens his mouth: "You actually have in your room a Diana, a Magnum Classic?" I turn my head. In fact I had completely forgotten. Peeking out from behind my red couch is the balance of an air rifle. Whether it is a Diana, I don't know. The gun is only in my room to prevent its descent into unauthorized hands. The boy though, he is fascinated. Knowing I have a Diana in my room loosens his tongue. He speaks about himself. It is the beginning of our long work together.

What does this snapshot of therapy have to do with ISAP's role in public space? The group we keep often influences our perceptions and thinking. Lively communities distinguish themselves by *discourse*. We communicate eagerly with one another through the media, at social events, in schools, at work, and in political spheres. We discuss, crack low jokes, muse, blame, brood, we become excited about and look forward to Roger Federer's offspring, climate warming, stress, banking secrecy, or the harbor crane at the Limmat River. Public discourse may be a minor exchange of contents but with it we calibrate attitudes and viewpoints. Without realizing it we attune to and adopt current modes of thinking and perceiving. That which doesn't belong to the current canons of opinion is not "in" and is hardly given a thought. The problem is, public discourse tends to one-sided and narrow thinking. It must, so that communities can come together and experience themselves as unities. It determines attitudes that are celebrated as correct. For instance, people were convinced for more than 80 years that "Jazz is the devil's music!" Pedagogues and psychologists admonished the schools to guard young people against the influence of this "Negro music," proving with scientific meticulousness that listening to this music led to degeneration of the brain. Public discourse imposed this perspective upon the population; "one" then thought accordingly.

What is the situation today? When we listen to the public discourse on psychology and psychotherapy, it is unequivocal: Belief in the formability of human behavior dominates. Thanks to efficient training programs, juvenile violence can be unlearned, and with the neurosciences we stand shortly before unlocking the secrets of the psyche! Human behavior is prognosticated with computer calculations, and with brain scans we can clearly register the causes of moods and feelings. The brain has morphed into a magic looking glass that reveals to us the secrets of schizophrenia, psychopathy, and depression. The key to the background of the human being is hidden in the grey matter. Measurable solutions, concrete goals and competencies are striven for and conveyed. The danger of this mode is that it reduces psychotherapy to a predictable process in which the goals and behavior targets are defined already in advance.

What does this have to do with ISAP? The public needs *counter voices*. It needs people representing attitudes that are at the moment not in vogue, but that indeed picture central and important aspects of human existence. It needs an institution that proceeds from a holistic picture and understands people as *thinking and spiritual beings*. An institution that dares to speak of soul and conceives every single psychotherapy session to be a curious, inquisitive exploration into the person's being. A psychology that draws on history, myth and art to make valid statements about people. It needs a school in which peoples' inner lives, dreams, fantasies and inspirations also have meaning.

Such psychology helps us in our work with people. Thanks to the psychology of C.G. Jung, the boy's behavior, his fascination for the air rifle, is attributed with meaning other than that which would be generally understood today. The boy is not only a potential perpetrator of violence who can be altered by anti-aggression training but also a person whose fascination with violence expresses a search for himself, his inwardness. The weapon is a symbol that allows him access to his feelings and fantasies. ISAP represents and disseminates a psychology thanks to which we can discern such expression of the psyche. It understands the soul as an autonomous power that expresses itself unconsciously not to prepare for an act of violence but because inner processes must present themselves and develop. ISAP makes a necessary and valuable contribution to the public discourse.

APPENDIX

ISAPZURICH LEADERSHIP 2004-2014

ISAPZURICH Council (formerly, Officers Committee)

President	Paul Brutsche	2004-2008
	Murray Stein	2008-2012
Co-President	Ursula Ulmer	2009-2010
	Marco Della Chiesa, Isabelle Meier	2012-present
Vice President	Stacy Wirth	2004-2008
	Ursula Ulmer	2009-2010
	Doris Lier	2010-2011
	Erhard Trittibach	2012-2013
Treasurer	Stefan Boëthius	2004-present
Secretary	Stacy Wirth	2009-2010
	Erhard Trittibach	2010-2012
Director of Program	Nathalie Baratoff	2004-2009
	Urs Mehlin	2009-2011
	Nathalie Baratoff	2011-present
Director of Studies	Katharina Casanova	2004-2008
	Christa Robinson	2008-2010
	U. Ulmer, Yuriko Sato (ad interim)	2010-2013
	Marianne Peier-Baer	2013-present
Director of Admissions	Doris Lier	2004-2009
	Monique Wulkan (March - June)	2009
	Ursula Ulmer (ad interim 2009)	2010-present
Director of Administration	Karen Evers	2004-2006
	Sandra Schnekenburger	2006-2010

Director of Operations	Karin Buchser	2009-2012
	Susanne Chapuis	2012-2014
Head of Administration	Begoña Martin Calzada	2014-present
	Chairs of Standing Committees	
Nominations	Ursula Hohler	2005-2010
	Christa Robinson	2010-present
Promotions	Jan Peter Hallmark	2004-2008
	Hannah Hadorn	2008-2012
	Lucienne Marguerat	2012-present
Counseling Service	Lucienne Marguerat	2004-2008
	Eileen Nemeth	2006-2008
	Vreni Bollag	2008-present
Jungian Odyssey	Cedrus Monte (Co-Chair)	2005-2007
	John Hill (Co-Chair)	2005-2011
	Isabelle Meier (Co-Chair)	2007-2011
	Stacy Wirth (Co-Chair)	2009-present
	Ursula Wirtz (Co-Chair)	2011-present
	Deborah Egger (Co-Chair)	2011-present
Zürich Lecture Series	Murray Stein, Nancy Cater (Spring)	2009-present
Academic	Douglas Whitcher	2008-2009
Departments	Nathalie Baratoff	2009-2011
	Antoinette Baker	2011 2013
	Kathrin Asper	2014-present
Care Group	Paul Brutsche	2011-2014
	Diane Cousineau Brutsche	2014-present
German Lang Task Force	Paul Brutsche (incl. Märztagung)	2009-present

	Special Services	
Library	Helga Kopecky, Librarian	2006-2014
	Andrew Fellows, Assistant	2009-present
	Eleonora Babejova, Assistant	2014-present
Studies Secretary	Franziska McSorley	2005-2006
	Elena Eckels	2004-2008
	Margaretha Jud (deceased 2011)	2008-2010
	Ana Frank	2010-present
Exam Coordinator and Exam Conference	Gine Ried-Hasler	2007-2008
	Yuriko Sato	2008-2012
	Christiana Ludwig	2012-present
US Student Loans	Nancy Krieger	2009-2012
Charta Delegates	Lucienne Marguerat	2005-2013
	Doris Lier, Michael Péus	2007-2010
	Katharina Casanova	2011-present
	Marianne Peier-Baer	2014-present
Social Events Coordinator	Katharina Casanova	2010-2012
	Gabriela Zahn	2012-present
Program Mailing	Ursula Kübler, Coordinator	2010-present
Ombudspersons	Maria Meyer-Grass	2006-present
	René Malamud	2005-2009
	Mario Jacoby (deceased 2011)	2009-2011
	Gary Hayes	2012-present

The foregoing is a good-faith attempt to record the names of individuals who have led ISAPZURICH since its founding in 2004 and through its 10th anniversary year in 2014. We apologize for errors and omissions and request your corrections. We acknowledge and thank all others, too many to name here, who have served through the years as committee members, special services, or both.

SOURCES – "ISABELLE MEIER INTERVIEWS PAUL BRUTSCHE"

No. 0: AGAP Business Meeting, Barcelona, Summary Minutes of Resolutions, August 30, 2004

AGAP
Association of Graduate Analytical Psychologists

AGAP Business Meeting
Fira Palace Hotel
Barcelona
Monday, August 30, 2004

SUMMARY MINUTES OF RESOLUTIONS

13:30 Lunch. President Deborah Egger called the meeting to order at 13:55.

1. Welcoming Address: The Members were welcomed and the President announced the overwhelming approval of the proposed new Constitution. As there are 636 members currently in AGAP, the required quorum of 314 Members participating by circular vote was met as follows: 392 in favor, 24 against, 2 abstaining.
It was announced that summary minutes and resolutions for this meeting would be recorded by Stacy Wirth in the absence of Secretary Yvonne Trüeb, who was ill. The President reviewed the rules for casting proxy votes, as allowed by the new constitution.
2. Election of Vote Counters was unanimously approved: Jan Bauer, John Granrose, Vicente de Moura, Dariane Pictet, and Marina Episcopi. The meeting's quorum was established as follows: For items on the Agenda allowing a vote by proxy, the quorum was 180. In this case, 91 votes constituted the simple majority required to approve all items

at this meeting. For agenda items not allowing proxy votes, the quorum was 111; a simple majority constituted in 56 votes.

The following Agenda items were voted on by a quorum of 180 Members:
3. The Report from the last meeting, in Cambridge 2001: approved unanimously.
4. The Report of the Honorary Secretary: approved by an overwhelming majority.
5. The Financial Reports for 2001, 2002 and 2003: approved by an overwhelming majority (179 in favor, 1 abstention).
6. The Executive Committee Members were unanimously released from liability.
7. The Annual Budgets for 2005, 2006 and 2007 were presented by Stefan Boëthius standing in for Yvonne Trüeb: Unanimously approved.

A motion had been sustained to delay this vote until after the vote on the proposed increase in dues. Stefan Boëthius was also foreseen to be AGAP's new treasurer should he be elected into the Executive Committee.

8. Proposed increase in dues from CHF 30 to CHF 50 per year: approved by an overwhelming majority (178 in favor, 2 abstentions).

The following Elections were conducted on the basis of the quorum of 111 (without proxies):
9. Election of the Executive Committee, Ethics Committee and Auditor (elections run by John Hill):

President: As the sole candidate for the presidency, Deborah Egger (Zürich) was re-elected by an overwhelming majority (102 in favor, 5 against, 4 abstentions).

Executive Committee: The following Members standing for election/re-election were unanimously approved: John Desteian (Minnesota), Diane Cousineau (Zürich), Stefan Boëthius (Zürich), Katharina Casanova (Zürich), Stacy Wirth (Zürich),

Mario Castello (Paris/Geneva), Dariane Pictet (London), Constance Steiner (Zürich).

Prior to the vote, the wish had been expressed to develop a more diversified membership in the ExCo (i.e., less Zürich-dominated). It was suggested that—given the necessity for the Committee to meet at least six times yearly and lacking funds to pay for travel costs— wider participation could be accommodated by conference calls and internet connection, etc. The Committee vowed to work toward this goal insofar as the budget allows the purchase of the required technology.

Ethics Committee: The following Members standing for election were unanimously approved: Elisabeth Martiny (Johannesburg), Jan Bauer (Montreal), Audrey Punnett (San Francisco), Craig Stephenson (Paris).

Auditor: The following Members standing for election were unanimously approved: Monique Wulkan (Zürich), Antoinette Baker (Zürich).

And in addition:

Outer Circle: The following Members standing for election/re-election were unanimously approved: Walter Boechat (Rio de Janeiro), Shirley Halliday (Vancouver), Toshio Kawai (Kyoto), Susan Pollard (Adelaide, Australia), Patricia Skar (Oxford), Ulla Olin Stridh (Lindingö, Sweden), Ursula Ulmer (Cape Town).

The following Agenda item was voted on with a new quorum of 87 (of 175 votes present) that was established to account for the fluctuation of Members attending the meeting at this point (including proxies):

10. The proposed delegation of training to a Zürich sub-group: approved by an overwhelming majority (149 in favor, 17 against, 9 abstaining).

Prior to the vote, the proposal was debated at length. For all the differences of opinion and feeling, the common ground was great sadness about the state affairs at the C. G. Jung Institute (Zürich) and about the necessity for AGAP to vote on this proposal.

After the vote the President noted the general consensus of those present: The AGAP delegated training should take every opportunity to come together again with the C. G. Jung Institute (Zürich) if ever a viable possibility presented itself in the future. This sentiment specifically is entered into the Minutes at the request of Members attending the meeting.

IAAP's President, Murray Stein, commented that AGAP's decision to open a training program, run on democratic principles by an IAAP Member Group, conforms to IAAP's regular standards and practices for Member Groups with Training. He emphasized, because AGAP itself is a democratic organization its members are empowered to call for another vote on the delegation of this training and, should it not meet with majority approval in the future, it can be changed or canceled.

Other:
> 11. Members willing to serve as additional Delegates to the IAAP Delegates Meeting on Wednesday, September 1, 2004, were asked to register with the President.
> 12. No miscellaneous items were presented.

Members were thanked for their participation and the meeting was closed at 16:45.

Delegates to the IAAP Meeting: Peter Ammann, Maria Teresa Alvarez, Christina Becker,
Mario Castello, Marina Conti, Diane Cousineau, John Desteian, Deborah Egger, Jim Fitzgerald, John Granrose, John Hill, Mary Anne Johnston, Carmen Martin, Charlotte Mathes, Christina Oldfelt, Dariane Pictet, Kusum Dhar Prahbu, Audrey Punnett, Aileen Campbell Nye, Gert Sauer, Jody Schlatter, Susan Schwartz, Patricia Skar, Constance Steiner, Craig Stephenson, Mae Stolte, Ursula Ulmer, Joanne Wieland-Burston, Luigi Zoja.

September 2, 2004
Approved by the Assembly at the Business Meeting, Cape Town, August 13, 2007.

Deborah Egger, President _____

Stacy Wirth, Acting Secretary _____

****

No. 1: Initiative Pledge Campaign: Results, May 17, 2004 (email)

Dr. Stefan Boëthius
Im Obstgarten 11
8700 Küsnacht

Zürich, May 17, 2004

To: The Members of the Curatorium, C. G. Jung Institute, Zürich
All Participants in the Pledge Campaign Initiative, All Colleagues in Switzerland (with email connections)

Pledge Campaign Initiative: Results

Dear Members of the Curatorium, Dear Colleagues,

On April 29, 2004, twenty analysts launched an independent initiative that invited people with close ties to the Institute to participate in a pledge campaign. The conditions were stated as follows:

"To support the Institute, I pledge to contribute funds under the condition that the present Curatorium withdraws no later than the end of the summer semester 2004 and is replaced by a new Curatorium previously constituted according to the results of a consultative election held by the community of analysts. The sum of pledged contributions will be reported to the Curatorium and the analytical community (without naming the individual donors) after May 14th, 2004."

I agreed to collect the pledges at my address and to inform the Curatorium and colleagues of the results. Accordingly: By Monday, May 17th, 2004 the pledged funds amounted to SFr. 340,000.
This total turns out to be much higher than expected. Many colleagues, pledging as each could afford, contributed to the gratifying result. It is impressive that such support materialized in less than two weeks – and this without having the pledges linked to any personal advantage. (By contrast: The policy enacted in October 2003 did little to promote the financial goal, although payment was linked to the advantages of professional affiliation with the Institute and the possibility to exercise training functions.)

The current result speaks for the widely held conviction that this Institute's chances of survival lie only in a new start with a new leadership gained by community consensus. The Institute could be saved with the promised funds. It is now up to the Curatorium to open to the available support or, rather, to act against the testified desire of the analysts' large majority. The hope is that the communal voice will be heard – and that the coming weeks will open the way to make use of the potential at hand to insure the Institute's survival.

In the name of the Initiative group, I extend hearty thanks to all who participated in solidarity and contributed to this campaign's impressive outcome. Should the conditions for new leadership not be met your pledges will be annulled on July 9, 2004 at 12:00AM. Everyone receiving this mail will be notified of the final upshot.

With friendly regards,

Stefan Boëthius

PS: This email has been sent out to the email address list of the Info-Forum. There are many colleagues who are not on this list. Please feel free to distribute this message to them.

****

No. 2: Development ISAPZURICH October 2004–February 2006

	per October 2004	per February 2006	Projected: per October 2006
Participating Analysts			
Category A (active)	66	78	
Category B (passive)	7	13	
Total	**73**	**91**	
Interested Analysts	34	65	65
Students			
Full-time Training Candidates	20	21	
Full-time Diploma Candidates	14	29	
Matriculated Auditors, Cont. Ed	3	9	
Matriculated Auditors, AJAJ	0	3	
Total	**37**	**62**	
Expected graduates (diploma)	0	4	
Auditors (Public Lectures)			
Demographics			
German-Speaking (includes bilingual)	5	19	
English-speaking	32	50	
Men	8	16	
Women	29	46	
Curriculum Offered			
No. of lectures			
No. of seminars			
Counseling Service			
Interviews w/ Diploma Candidates	0	19	
Client Interviews	0	5 (+4 from abroad)	
Referrals	0	3	

Budget: Balanced Annual operating expenses covered (tuition & analysts' participation fees)	yes yes	yes yes	yes yes
Secretarial Staff	0%	1 position, 100%	1 position, 100%
Honoraria paid (Training & Committee Functions)		2004-2005 50% of standard	2005-2006 50% of standard
Library Development			
Student Loans			
Legal Fund			
Unstipulated			
Total	**CHF 260,000** (per Dec 04)	**CHF 520,000**	

****

No. 3: AGAP Meeting for the Constitution of the Delegated Training Program in Analytical Psychology, Erlenbach/Zürich, Erlengut

AGAP Association of Graduates in Analytical Psychology

September 9, 2004

Summary Minutes / Election Results

1. Opening of the Meeting
AGAP President Deborah Egger opens the meeting at 19:40. Stacy Wirth agrees to take the Minutes. The President briefly reviews the results of the AGAP Business Meeting in Barcelona that led to tonight's meeting. She clarifies that membership in the new Training Program is open to AGAP members only; applications for AGAP membership can be submitted immediately.

The President explains that, for purposes of this meeting, membership and voting rights in the new Training Program will be established as follows: Those who wish to join tonight and to have voting rights must sign and submit the provided form, which confirms the agreement to pay the proposed dues. This signature, in turn, confirms the member's agreement to pay the exact amount of dues as determined by vote in the course of this meeting. As a member's form is signed and submitted, he or she will be given a green voting card. A second form is available for analysts who prefer not to obtain membership tonight—but who wish to remain informed about future developments; such people, as well as analysts not attending this meeting, are invited to apply for membership any time in the future.

Concern from the floor is immediately expressed about the fact that some AGAP members did not receive the invitation to this meeting. The President acknowledges and apologizes for the oversight. She explains that these are partly attributable to a shortage of good administrative support, due to illness in the secretariat. The President asks us to tolerate the imperfections of this fledgling undertaking. She urges us to work together for resolution as soon as possible.

With regard to concerns about tonight's elections, the President reminds us that (1) the broadest possible contact was sought by invitation through the Info-Forum. Via the Info-Forum, AGAP members and other analysts in the region were encouraged to submit nominations for tonight's elections. (2) At tonight's meeting, all nominations from the floor will be entertained. (3) Officers and committee members elected in this meeting are transitional and will be voted upon again in the coming year.

By general consensus, it is agreed that members with special financial circumstances will be allowed to request a waiver either from the requirement to pay, or a reduction of, the established dues.

The President hands the Meeting over to Urs Mehlin and John Hill, who agreed to conduct tonight's voting. Urs Mehlin submits the list of analysts who formally excused themselves from this meeting.

Urs Mehlin recalls the meeting held at Erlengut on September 13, 2003 at which it was urgently recommended that a solution to the crisis at the Jung Institute be found. This meeting led to AGAP's delegation of this new training program in Zürich. He thanks Deborah Egger, President of AGAP, along with other colleagues who worked during the foregoing months to make a new training program possible. With apologies for those unintentionally not mentioned: Antoinette Baker, Nathalie Baratoff, Susanne Boëthius, Stefan Boëthius, Paul Brutsche, Katharina Casanova, Diane Cousineau-Brutsche, Karen Evers, Doris Lier, Kari Lothe, Jan Peter Hallmark, John Hill, Hans-Peter Kuhn, Sandra Schnekenburger, Constance Steiner-Blake, Stacy Wirth.

2. Report on the AGAP Meeting in Barcelona (John Hill)

Regarding the adoption of the new constitution, as AGAP presently numbers 636 members, the required majority to approve the new Constitution according to Art. IV of the old Constitution is 314 (recte 319) votes (majority of all members) or two thirds of all those voting, whichever is greater. The result of the written vote was as follows: 392 yes, 24 no, and 2 abstentions. The new constitution was, therefore, validly adopted. Prior to the vote, the proposed delegation of training to a Zürich subgroup was long debated, causing the meeting to run overtime. The delegation of training was finally approved by an overwhelming majority of voting members present together with members voting by proxy (149 yes, 17 no, 9 abstentions).

There was great common concern that the AGAP delegated training should work toward coming together again with the C.G. Jung Institute Zürich as the viable possibilities present themselves in the future. Following John Hill's remarks, the President reports that, after the AGAP Barcelona meeting, other Jungian schools indicated firm interest to build up an exchange and a collaboration with the new Zürich training program.

3. Establishment of Membership Dues and Voting Rights/Election of Vote Counters (Urs Mehlin)
A motion is overwhelmingly approved to take a consultative vote on the dues proposed for members who will participate in the new training as follows: Category A: CHF 600; Category B: CHF 1000. The proposed dues are approved by an overwhelming majority with the reservation that application for full waiver or dues reduction can be made.
The vote counters are unanimously approved: Ruth Ledergerber, Katharina Schmid, Nathalie Baratoff.
The submission of the signed agreements indicates 53 voting Members are present. The quorum is therefore established at 53 with a simple majority of 27 votes required to approve motions and accept elections at this meeting.
The members are informed of their right to call for secret ballots. At the same time they are encouraged to proceed, if possible, without calling for this right. Time will be saved—and moreover, transparency of communication will be upheld.

4. Announcement of Nominees (Urs Mehlin)
Urs Mehlin initially eliminates nominees who had been named without their explicit consent. For all offices, he also takes nominations from the floor. (See below.)

Prior to voting, the nominees present and running for the Officers' Committee introduce themselves and take questions from the floor. It is generally acknowledged that, while many have become familiar with and have performed according to the demands entailed in the respective offices, experience remains to be gained.

5. Election of Offices and Committees (Urs Mehlin, John Hill)
<u>Candidates for President</u> eliminated as non-consenting presidential nominees: Ursula Hohler, Peter Ammann, Elynor Barz, Elisabeth Hartung, Linda Briendl, Irene Bischoff; no nominations from the floor. Contending: Paul Brutsche and Karen Evers.
The election of the President is strongly debated. Opinion divides along two basic lines: (1) One maintains that, with all due respect for Paul Brutsche, his presidency could be encumbered by the history of his involvement with the present Curatorium; Karen Evers would therefore be desirable. (2) The other maintains that, in this difficult start-up phase, we urgently require a president who would assume office with experience, in this case, Paul Brutsche would be preferred.

Prior to the vote, we are reminded that these elections will lead to a transitional leadership and that elections will be conducted again in one year.

President: Paul Brutsche is elected with 34 votes.

Vice President: Stacy Wirth, contending with Karen Evers, is nominated from the floor. Stacy Wirth is elected with 43 votes.

Treasurer: Stefan Boëthius, singular nominee, unanimously elected. (2 abstentions)
Director of Administration: Karen Evers, singular nominee unanimously elected.

Director of Program: Nathalie Baratoff, contending with Ian Baker, is nominated from the floor. Nathalie Baratoff is elected with 34 votes.

Director of Studies: Katharina Casanova, singular nominee, elected with 46 votes.

Director of Selection: Doris Lier, singular nominee, elected with 50 votes.

Selection Committee: Katharina Schmid withdraws her nomination. Diane Cousineau- Brutsche, Jan Peter Hallmark, Tonie Baker, and Ursula Wirtz, contending with nominations from the floor: Jody Schlatter, Marco Della Chiesa, Monique Wulkan, Peter Ammann.
Voting results: Diane Cousineau-Brutsche, 45 votes; Jan Peter Hallmark, 30votes; Tonie Baker, 39votes; Ursula Wirtz, 38votes; Jody Schlatter, xx votes; Peter Ammann, 36votes; Monique Wulkan, 31votes; Marco Della Chiesa, 28votes.

The motion to discuss this outcome and eventually call for a re-vote is approved.

Discussion: The common consensus is that there are too many Selection Committee Members for the projected number of students – and too many women. Monique Wulkan volunteers to withdraw because, among the women, she had received the least votes. Accordingly, the motion is approved to withdraw the man who had received the least votes. The motion is approved to re-vote on the group of nominees as follows:

Diane Cousineau-Brutsche, Jan Peter Hallmark, Antoinette Baker, Ursula Wirtz, Jody Schlatter; Peter Ammann: elected, as a group.

Training Committee: Cedrus Monte withdraws her nomination. Katharina Schmid, Gary Hayes, Sandra Schnekenburger are each unanimously elected.

Program Committee: Diane Cousineau Brutsche's nomination is withdrawn due to her election to the Selection Committee. Maria Bernasconi, Ursula Lenz-Bücker, Susanne Boëthius, John Hill, Urs Mehlin contending with nominations from floor, Cedrus Monte, Ian Baker, Bernhard Sartorius, Brigitte Egger: a group, unanimously elected.

6. Questions and Adjourning of the Meeting
No questions are brought to floor. Urs Mehlin underscores the need for help in many areas, not least in the functions of a secretariat, which we cannot yet afford to employ. Additionally, the analysts present are urged to contact colleagues who missed this meeting and to advise them to get information from the home page at www.isap.info. Alternatively, they can contact Stefan Boëthius.

The President adjourns the meeting at 22:55.
For analysts attending the meeting and those excused, see the next page.

Approved, September 28, 2004

Deborah Egger, President
(signed)

Stacy Wirth, Acting Secretary
(signed)

For information on the new program and membership contact:

www.isap.info
or
Stefan Boëthius Tel: 079 401 32 37
Im Obstgarten 11 Fax: 01 854 74 49
CH-8700 Küsnacht Email: st.boethius@telia.com
Erlengut Meeting, 9.9.04

Analysts attending:		**Formally excused:**
Ammann Peter	Malamud René	Aeschbach M.-l.
Asper Kathrin	Marguerat Lucienne	Ammermann Christof
Baker Antoinette	Mehlin Urs	Bachetta Florence
Baker Ian	Meyer Christina	Barz Ellynor
Baratoff Nathalie	Monte Cedrus	Barz Helmut
Bernasconi Maria Anne	Moser Annemarie	Briendl Linda
Blum Denise	Nemeth Eileen	Brutsche Paul
Boëthius Stefan	Pèus Michael	Casanova Katharina
Boëthius Susanne	Ried Hasler Georgine	Cousineau-Brutsche Diane
Bürge Max	Sartorius Bernard	della Chiesa Marco
Dömer Michael	Schlatter Jody	Dedola Rosanna
Egger Brigitte	Schmid Katharina	Funkhouser Arthur
Egger-Biniores Deborah	Schmid Maxine	Guggenbühl Allan
Evers Karen	Schnekenburger Sandy	Gyurina Dennis
Hadorn Hanna	Schoeller Rütger	Hegge M.
Hallmark Jan Peter	Smith Joan Allen	Locher Roberta
Hartung Maria-Elisabeth	Steiner-Blake Constance	Müller Marianne
Hayes Gary	Thommen Nicole	Rohland Johannes
Hill John	Vogelsang Ethel	Scheidegger-Jans F.-X.
Hohler Ursula	von Uslar Ilsabe	Stein Murray
Jacoby Mario	Weis Heike	Scategni Wilma
Kapferer Gudrun	Whitcher Douglas	Stüssi Franziska
Kaufmann Rolf	Wirth Stacy	
Kennedy Emmanuel	Wirtz Ursula	
Koch Monique Wulkan	Zahn Gabriela	
Körner Waltraud	Zavala José	
Kuhn Hans-Peter		
Ledergerber Ruth		
Lenz-Bücker Ursula		
Lier Doris		
Lothe Kari		
Ma Shirley		

No. 4: Letter to the Curatorium, April 3, 2005 (email)
ISAP Zurich Internationales Seminar für Analytische Psychologie Zürich
International School of Analytical Psychology Zurich
AGAP Post-Graduate Jungian Training
Hochstrasse 38 CH-8044 Zürich Tel 043 344 00 66 Fax 043 268 56 19

TO: The Curatorium of the C.G. Jung Institute, Zurich
CC: Students and Analysts of ISAP, Zurich, Interested Analysts of ISAP Zurich
RE: Statement on recent Curatorium Letters

April 3, 2005

Dear Members of the Curatorium,

In two recent letters, one of February 25th addressed to students, analysts, and patrons of the CGJI and another one written personally to ISAP students who have ex-matriculated from the Jung Institute, you refer to the case filed against AGAP by Brigitte Spillmann, Ernst Spengler, and Robert Strubel claiming that, if the case is successful, individuals holding an ISAP diploma will be "unable to qualify for IAAP membership."

It is obvious with this undignified tactic of intimidation that you want to spread uncertainty among the students in the hope of winning them back to the Institute. We take strong exception to such behaviour, which requires us to make the following statements:

1. The Legal Case. As we have had confirmed by the IAAP, upon the successful completion of the ISAP training and receipt of the Diploma, a graduate will be admitted into AGAP. By virtue of membership in AGAP, one is automatically enrolled in the membership of the IAAP. The ISAP Training in Zurich was approved by a large majority of members at the AGAP Business Meeting in Barcelona last August (149 for, 17 against). The adoption of the new AGAP Constitution, which was confirmed by written ballot (as required by the old constitution), also enjoyed an overwhelming majority vote of those participating: 391 yes to 6 no votes.

By challenging both decisions, the three plaintiffs want to prevent AGAP from offering its own training. Their argument for this is that AGAP is not an association for the promotion of professional interests of Jungian analysts but exclusively an alumni association of the Jung Institute. It therefore may not offer its own training program unless the purposes of the association are changed by a unanimous vote of the AGAP members. It is not the place here to disprove that farfetched argument, for that will happen in the course of the legal proceedings. We simply want to point out that we consider this way of doing as a further manifestation of the Curatorium's destructive attitude and blatant disregard of a clearly expressed majority will.

As an IAAP constituent Society, AGAP is mandated to look after the interests of its trainees and will continue to safeguard those interests, including ensured IAAP membership, through our work in close association with the IAAP. In the highly unlikely event that the complaint would succeed in removing the ISAP training from within AGAP, there are avenues available to ensure the professional safety of our trainees and to guarantee that anyone holding an ISAP Diploma will be duly recognized by the IAAP. These possibilities will be activated, if proven necessary by the results of the legal process.

2. <u>The Professional Situation.</u> To your comment that the ISAP Diploma "neither entitles one to practice his or her profession in Switzerland (and hence, neither elsewhere)," we stress the following:

The training at ISAP is currently geared to international students. This includes Swiss residents who are either already qualified for the psychotherapy practice license or for those who want to work with persons interested in the development of the soul's process without medical insurance coverage. The program, as such, is in no way different from that for Swiss trainees who study at the CGJI under the "i" regulations. Our diploma qualifies an individual as a Jungian Analyst worldwide. Moreover, graduates of ISAP are qualified for SGAP membership under the same conditions that hold for graduates of the CGJI.

As a part of our ongoing development we have been considering applying for membership in the Swiss Charta, in order to gain recognition from the Swiss Department of Health. ISAP Participant analysts just recently voted unanimously to approve the decision to apply for Charta membership.

3. <u>The Quality of ISAP Training.</u> We take issue with your understanding of "quality" in analytic training, in that you seem to link it only to purely formal and legal recognition through Institutions and Government agencies. While we obviously recognize the importance of such recognition, the measurement of real quality in training is determined in other ways – in the fullness and level of the teaching program (lectures, seminars, colloquia), in the solidity of

the training curriculum, in the quality of the trainers, and in the organizational and human setting, which creates a beneficial training atmosphere. ISAP offers quality in all these areas. Perhaps it would be wiser, rather than worrying about ISAP as "a qualitatively insufficient Jungian degree in the Zurich area," to ask yourselves if the quality of your own English language training program is still up to par after the mass exodus from the Jung Institute of most of the best teachers and analysts in Switzerland.

4. Allegations Need to be Substantiated. In closing, we insist that you stop making false and misleading claims, including those in connection to ISAP participants who have reneged on their participation in ISAP to go back to the Jung Institute. We do not know of even one such person. This futile hope of yours for an obviously missing movement is another way that you simply try to create an atmosphere of insecurity.

Sincerely yours,

Deborah Egger, President AGAP   Dr. Paul Brutsche, President ISAP

****

No. 5: Christian Gaillard, IAAP President, Letter to Mr. John Betts, September 13, 2005

INTERNATIONAL ASSOCIATION FOR ANALYTICAL PSYCHOLOGY
INTERNATIONALE GESELLSCHAFT FÜR ANALYTISCHE PSYCHOLOGIE
ASSOCIATION INTERNATIONALE DE PSYCHOLOGIE ANALYTIQUE
ASSOCIAZIONE INTERNAZIONALE DI PSICOLOGIA ANALITICA
ASOCIACIÓN INTERNACIONAL DE PSICOLOGÍA ANALÍTICA

Secretariat: Office of the President:
P.O. Box 115 9 rue de la Boulie
CH-8042 Zürich, Switzerland Verrières-le-Buisson
Tel. 41-1-261-3393 France 91370
Fax. 41-1-272-9606 Tel/Fax (331) 69 53 15 58
iaap@swissonline.ch, email: president@iaap.org

Mister John Betts
1190A Fort Street
Victoria, BCV8V 3K8
Canada
jbetts5@telus.net

Dear Mr. Betts,

In answer to your e-mail of March 22 addressed to our Zurich secretariat, I have the pleasure of confirming that AGAP is one of IAAP's Constituent Societies, designated as such at the Constitutional meeting held in 1956. It is recognized as a Group Member of the IAAP with Training Privileges. It is my understanding that AGAP exercises this training privilege through the International School for Analytical Psychology (ISAP), and that, upon successful completion of the ISAP training and receipt of the Diploma, a graduate is admitted into AGAP. According to IAAP's Constitution (Art. 3 A 1.), a member of one of IAAP's Group Members automatically becomes a member of IAAP. I hope this answers your questions clearly. You may, of course, make this information available to those around you who perhaps have similar questions.

With my best regards,

Prof. Dr. Christian Gaillard,
IAAP President

No. 6: ISAPZURICH: ANNUAL REPORT 2004-2005

1. **OFFICERS COMMITTEE**
   From the standpoint of the Officers Committee, this first year of ISAP's existence was completely under the sign of start-up and development: The launching of this new training program has been by nature a very demanding and work-intensive job – if also an extraordinarily satisfying and motivating one, because so much is given back. In no time at all and out of nothing, we were faced with the challenge of bringing forth an institution that met the complex requirements of any Jungian training program. In the course, we were repeatedly confronted with the need for quick decisions in the greatest imaginable variety of areas. The consequent demand for constant movement and strength of resolve – admittedly, not always part of the therapist's natural equipment – kept us constantly learning.

   1.1 **The Complexities of the Job**
   To characterize the complexity of the job with a few key words, the following items either had to be found, realized or accomplished: a suitable building; an acceptable name for the institution; training regulations for international students and a Charta version for the Swiss; a program of lectures & seminars for the first winter semester; advice on the subject of a children's program; legal procedures with the immigration office to secure visas for international students; the establishment of a secretariat and employment of personnel; development of a bookkeeping system; the installation and maintenance of a web site; the launching of advertising and fund-raising drives; translation of all materials in German and English (Testatheft, registration forms, exam result forms, guidelines for supervised case work, ethics guidelines, a diploma); the production of an ISAP-logo and brochure as well as flyers for the Counseling Service; coordination with the team responsible for the planning of next year's Jungian Odyssey; build-up of a library; application for Charta membership, etc.

   1.2 **Officers Committee Meetings**
   To manage these and many other urgent jobs, the OC held 33 regular meetings in the course of this fiscal year always on Wednesday evenings from 7:00 – 11:00 PM.
   To gain time for the discussion of fundamental orientation, the OC also went on two weekend retreats from Friday through Sunday. The first took place at Haus St. Dorothea in Flüeli-Ranft and the second at Eranos, Ascona.

   1.3 **Liaison with the AGAP ExCo**
   Also in the course of the year, members of the OC participated in four different weekend meetings with the AGAP ExCo. In these sessions we found the ExCo members eager to hear about ISAP's progress, ready to provide constructive criticism and support where necessary, and prepared to report

on ISAP to AGAP members at large. In the same context, joint sessions with Marianne Müller furthered the purpose of ISAP liaison with the SGAP.

1.4 Contact with Students

The OC met at regular intervals with the two representatives of the Student Association. We were thereby informed about student concerns and able to appropriately respond to these.

In addition, there were two meetings of the OC with the whole student body, which provided the chance for open questions and the mutual exchange of concerns and ideas.

1.5 ISAP Meetings

In the first year, four ISAP meetings were held for participants and interested analysts. At these well-attended events, we began to practice the still unaccustomed democratic freedoms of co-creating our new program, co-determining its fate, and mutually carrying the responsibility – and all of this with evident pleasure.

1.6 Down to Basics

ISAP's first year can be counted as a very successful and promising one. This has been, above all, thanks to the strong and palpable support of many colleagues who committed themselves with enthusiasm and selflessness as instructors, helpers, active committee members and generous donors – among the latter we must count also friends from the outside, the larger part of the donations, however, came from our colleagues.

Yet we cannot avoid the reality that much developmental work remains – and we must simultaneously ask how we can insure that, in the future, the burden on the colleagues involved remains within reasonable limits.

Paul Brutsche, President

2. ADMINISTRATION

ISAP opened its doors in October 2004 and immediately began a full schedule of classes and seminars. The organization of the front office and the logistics of managing the classes at that time were all done by volunteers. Despite the enthusiasm and high level of competence of the student and analyst volunteers, it quickly became apparent we needed a professional secretariat. The decision was made to hire an experienced secretary for a 60% position to supplement the 10% position filled by the Director of Admissions/Director of Studies secretary. We were very fortunate that Elena Eckels (whose long experience is invaluable) agreed to take charge of the admissions secretarial work and to find another very experienced and capable person, Franziska McSorely, who took her position in March 2005. After the end of the summer semester 2005, it was again apparent that, as the student body grew and the demands of the secretariat grew, we needed more capacity. The Officers Committee decided, therefore, to hire Karen Evers for a 20% position to work in the secretariat (due to the fact that she did not need any training and would be able to jump in immediately). This provides the secretariat an 80% position. Karen Evers functions in her position as Director of Administration in times outside this 20% position. In addition, we have been able to rely on the caretaker, Mr. Lee, to assist with opening and closing the main door in the evening and on weekends.

The secretariat began with a donated computer and a laptop. Once we had a professional secretary who took over the bookkeeping and other office duties, it was clear we needed a proper computer, and so we purchased a Dell desktop computer. In the past year, we have also made a contract for a copier machine that is used for office work for routine copies and printing, as well as for producing all our programs and directories. Students are assigned their own code for making copies so that we are able to keep track and bill the individual students for the copies they make.

The office system initially was not optimally networked and computers crashing or failing to log onto the Internet was a problem. We were able to find a reliable and competent IT firm to organize our network. At present, we are reviewing bids by two IT firms to reconfigure the office network to include a secure firewall, an offsite data storage system (for security and backup) and to integrate into the system our planned library software.

Other activities: We have a system in place now for collecting entrance fees for lectures outside office hours using a small number of students. Also, we began in SS 2005 to submit routinely open lectures titles to the NZZ and the Tages Anzeiger for inclusion in their weekly listings of Zürich activities. We have begun hosting the Jungian seminars group twice/year.

2.1 Opening Hours

Our secretariat is occupied from 8:30 to 16:30 every day (closed for lunch usually 12:30 to 13:30). Office hours for students are reduced in order to allow the secretaries time to work uninterrupted. These hours are posted on the main office door.

Karen Evers, Director of Administration

3. ADMISSIONS COMMITTEE

The Admissions Committee is made up of six members – currently one man and five women. We managed a total of 216 interviews and another 14 with students requesting promotion in the training program. All in all, the interviews were very satisfactory. We noted that our students are engaged and take their training very seriously.

Apart from the interviews we had five meetings in this first year; the main theme was the students themselves. Further important tasks: We reflected on the manner in which we work together and we designed several new forms;

we had to outline our strategy with respect to granting exceptions to the study regulations.

Above all in the first semester with the large number of applications our work was very intensive and open – due to which we quickly found ourselves to be a well-functioning committee.

Doris Lier, Director of Admissions Committee

4. TRAINING COMMITTEE

One year ago we began our first ISAP semester and had hoped for about 20 students. About the end of November we had finally registered 37 students, who all wanted to be individually advised and admitted to training. As is generally known, the majority was in the English program; from the beginning, however, there were also four German-speaking candidates in the program. Many of the students who transferred from the C.G. Jung Institute wanted to take exams with analysts they knew. Therefore in the winter semester 2004/05 there were fifty-nine (59) exams given, in the following semester forty-one (41) for a total of 100 exams this past year – 7 of which were case exams.

In the summer semester 2005 we increased to fifty-four (54) students and now in the current winter semester 2005/06, our second year at ISAP, we have enrolled 63 candidates and auditors. Since December 2004, Elena Eckels has been employed to work for the Director of Studies. Without her knowledge and ability, it would not have been possible to manage the enormous work.

Semester	Anzahl Stud	weibl.	männl	Diplom-kand.	Ausbild.-kand.	Imm. Hörer	AJAJ	Sprache
WS 04/05	37	29	8	14	20	3		32 E, 5 D
SS 05	54	40	14	25	21	6	2	47 E, 11 D*
WS 05/06	62	46	16	29	21	9	3	50 E, 19 D*

*Some bilingual

Some of the auditors are working toward applying to the training program or to complete the certificate program. We expect we will be able to award the first diplomas at the end of next semester. Right now, there are 8 active applications for summer semester 2006. We will no longer be so overrun as at the beginning of ISAP, which allows us to proceed at a calmer pace.

Lucienne Marguerat has agreed to take over the leadership of the Counseling Service. In innumerable meetings with the diploma candidates, and with Sandy Schnekenburger, she has been responsible for the creation of a beautiful Counseling Service flyer and the placement of English language advertisements. She has also planned further steps to publicize possibilities for client referral to our diploma candidates.

In the Jungian tradition, students from all over the world come to Zurich: in first place is the USA with 16 students, followed by Canada, Japan, Germany, and Switzerland. Other students come from Italy, Denmark, Norway, France, England, Belgium, Brazil, India, Sweden, South Africa und Venezuela.

With that I would like to raise an issue that comes with the high number of foreign students: many of the students have problems finding analysands here in Switzerland in a reasonable time period, or they are already psychotherapists and have their clients in their practice at home. Again and again, I am asked in my capacity as the Director of Studies to find an individual solution for these candidates, which would allow them to spend only a part of the semester in Zurich, without compromising the quality of their education.
Katharina Casanova, Director of Studies

5. COUNSELING SERVICE
5.1 Meetings with Diploma Candidates

As the Counseling Service began its services, the Director met several times with the Diploma candidates. The purpose of these meetings was to get to know each other and also for the Director to learn of the questions and problems of the candidates. The candidates were also very helpful in establishing an advertisement plan for the Counseling Service, which included gathering names of different English language newspapers and clubs and creating diverse texts in English for the Internet site, for ads in English language press, and for the flyer.

Between November 2004 and June 2005, eight meetings took place at ISAP with many email exchanges in between. Since the creation of the flyer, contact with the students has returned to normal and is conducted mainly by telephone.

I would like to express my heartfelt thanks to the candidates who participated, for their patience and generous willingness to help.

5.2 Diploma Candidate Statistics

In the beginning there were seven, now increased to twelve diploma candidates, requesting client referrals and taking part in the Counseling Service. Eleven of these live in the Zurich area and one lives in Basel. Seven candidates have English as their mother tongue and are American or Canadian (1). Two are from Switzerland. The other four speak Spanish, Portuguese, Brazilian and Danish originally, but are also fluent in English.

5.3 Advertisement Campaign for the Counseling Service

At first a candidate suggested advertising on the internet site XpatXchange.ch. In the meantime, the Counseling Service is also advertising its services on the Homepage of ISAP in English and in German.

There are 2,500 flyers or "information cards" at ISAP, which advertise the Counseling Service in English and in German. These cards are available for Colleagues to make the Counseling Service known. They also serve to give the Candidates "backing" when they offer their services within their personal spheres or among ex-patriots of their various countries of origin. From the initiation of the Counseling Service and running until summer 2006, an ad appears in seven English language newspapers in the Zurich area and in Basel (1). Advertising is also planned in the local press (NZZ, Züritipp and Volkshochschule). In addition, clubs and, if possible, international schools will be visited and informed of our services by the Director accompanied by Stacy Wirth or Sandy Schnekenburger.

I cannot thank enough, Marcus Baker, our House Designer and Photographer, for his generous professional help in creating the flyer and ads with tremendous style!

5.4 Inquiries

Between June and October 2005, five persons consulted the Counseling Service and were referred to a Diploma candidate. Three of these came to the Service via the Internet, two were recommended to us from friends, and one saw our ad in an English language newspaper. Two persons did not reply to our response to their inquiry.

In the same time frame, four other inquiries were received from foreign countries or more distant regions. All of these requests, which came from Geneva, Portugal, Los Angeles (California) and Dallas (Texas), could be referred to colleagues in the local vicinity.

Lucienne Marguerat, Director of the Counseling Service

6. PROGRAM COMMITTEE
6.1 Program

The Program Committee (PC), consisting of nine analysts and three student representatives and led by the Director of Program, met nine times in this first year of ISAP's existence.

The committee's work included such tasks as writing minutes, proofreading the program, translating letters and drafting flyers for special occasions.

A special "Lecturer's Information" form was composed and sent to all lecturers, seminar, and colloquium leaders with the purpose of creating a pool of information not only for the use of the PC but also for students wishing to acquaint themselves with the fields of expertise and interests of the teaching staff. A folder with this information will ultimately be kept in the library.

In order to present a clearer overview of courses as well as to facilitate preparation for examinations, the semester program is now divided into subject areas based on examination topics. In WS 2005/06 the subject area "Child Psychology and Child Therapy" was added to their number.

The Jungian Odyssey, ISAP's summer intensive program, is presently being coordinated by a subcommittee under the leadership of Cedrus Monte. Although this function was originally a part of the PC, it was provisionally decided at a joint meeting of the OC and the JO sub-committees that the subcommittee should operate under its own organization and budget and be directly responsible to the OC. The Jungian Odyssey 2006 is planned to take place in Flüeli-Ranft with a welcoming party ahead of time at ISAP. Later evaluation will help determine how this committee might best fit within our governing structures.

6.2 Library

Over ten analysts and students have been involved in setting up the library in this past year. A wish list in both German and English was made available on our website. In the first nine months, donated books were received and ordered into provisional categories; a few books were bought.

A computer program was found that answered to our needs and the necessary software was purchased.

Book shelves went up in August, turning the storehouse of banana boxes into a (potential) library. Beginning in September, the acquisition of books became a more active phase, one of searching and buying, mainly from Internet sources. Four analysts are presently devoted to this task and making very good progress.

It is anticipated that the books will soon be ready for cataloguing. After this, the library will be ready for use. There is, however, still a need for additional volunteers for cataloguing and for library duty.

The library committee wishes to express its gratitude for all donations, both monetary or in the form of books and hopes that ISAP will soon have a library for our students to use and for all of us to rejoice in.

Nathalie Baratoff, Director of Program

7. **FINANCES, DONATIONS, ADVERTISING, PR, INTERNET**

Thanks to the selfless engagement of many colleagues – and to the unexpectedly high rate of student enrollment – ISAP's first fiscal year closed as per September 30, 2005 with a bank balance of CHF 714,392.10. CHF 165,000.00 was the original sum budgeted. This is a healthy financial basis for ISAP's build-up phase. The sum of donations for this fiscal year was CHF 519,656.00. Expenditures reached a total of CHF 171,513.02, while the original amount budgeted was CHF 163,800.00.

Communications investments went specifically to a website, a brochure, advertising in the print media, and for a flyer and internet advertising for the Counseling Service. The ISAP website was visited about 2,500 times per month. For these measures we spent only CHF 2,750.00. Print materials have been generated primarily in the fiscal year 2005/06; their cost will correspondingly appear in that period.

The most important PR measure has been the weekly listing of ISAP's public lectures in the culture calendars, "NZZ TICKET" and "TAGI ZÜRITIPP."

Stefan Boëthius, Treasurer

8. **DESIGN OF PRINT MATERIALS**

Since October 2004, the OC has been working with Marcus Baker to evolve a design for ISAP's print materials. The goal has been to come up with a logo and a basic layout that are recognizable. In the process, great care has been taken to avoid replicating designs that could be associated with the CGJI. This is because good marketing dictates the need to project ISAP's distinct and separate identity. Not least, materials that liken CGJI designs would stand to provoke justified legal claims that ISAP is attempting to capitalize on the CGJI "label."

All of Markus Baker's work through October 2005 was done for ISAP on a pro bono basis. This has included, among other things: the layout of the regulations for submission to the immigration office; Testatheft (certification booklet) & student ID card; four logo designs; the card and media ads for the Counseling Service; the ISAP brochure, enclosures & envelope. From November 2005 onward, Marcus Baker will invoice his work, but, in doing so, he will continue, generously, to consider our financial situation.

Logo: A provisional logo emerged in connection with the production of our first brochure. Until the finances improve, the red-black version would have to be substituted with the black-black as it appeared on the enclosures to the brochure. The OC is pleased with the design but, before adopting it permanently, wishes a green light from ISAP participants.

Brochures: ISAP's first brochure is intended to make ISAP known to prospective students and donors. An anonymous donor and partial sponsorship from Archives Printing enabled the full-color production. The brochure, as such, with a printing run limited to 2000 pieces, should be held as an experiment. The OC urges ISAP participants to give feedback on the design as well as on the content. Design of the brochure for the Jungian Odyssey on the other hand was done in-house in collaboration between the OC and the Odyssey committee. It is hoped that in the future finances will allow professional production. It needs be said here that the creation of the Odyssey website is also thanks to professional work of John Farr, who labored at practically no cost to ISAP.

Forms: All of ISAP's forms are currently being revised at the hand of Franziska McSorley. For the short-term this involves updating specific information and procedures. In the long-term, revision will require incorporating a logo and layout as mentioned above. Marcus Baker plans to create a database that will enable us to easily update old forms and also create new ones, as necessary. Parallel to all of this, we have experimented with a downloadable form that can be filled out on the computer. We are informed that this system has glitches; it remains to be seen whether these can be fixed or whether we will revert to forms that must be filled out by hand.

Regulations: In collaboration with the Admissions Committee, the OC is in the process of amending the international training regulations. This will entail changes to both content and layout. For the purpose of application for Charta membership, the OC has created a version, Charta Regulations. These were printed in a completely new layout consistent with the design that has been evolving in collaboration with Marcus Baker.
Stacy Wirth, Vice President

UPDATE ON THE LAWSUIT AGAINST AGAP

In June 2004, the CGJI Curatorium wrote to AGAP members worldwide announcing that a legal attack had been prepared should the membership approve AGAP's new Constitution. In October 2004 this threat was followed up by Curatorium President Brigitte Spillmann, Curatorium member Ernst Spengler and Robert Strubel (Head of the Selection Committee), who acted jointly as AGAP members to file a lawsuit against our founding association. Legal proceedings went into full gear when an obligatory meeting with the justice of the peace in November 2004 failed to bring resolution. Between

February 2005 and the end of October 2005, there followed the orderly exchange of legal briefs.

In the course, AGAP's trial attorney diligently examined the numerous claims and concluded that none of them hold up under the applicable law (Swiss Association law or *Vereinsrecht*). Indeed, he determined that all the proposed constitutional changes were within the parameters of AGAP's founding purpose and that all related voting procedures were correct. He argued moreover that – given the recent history of relations with the Curatorium – the Executive Committee acted in line with AGAP's duty to protect the members' professional interests when it presented the Barcelona motion to delegate training rights to a Zurich sub-group.

The final judgement was pronounced recently, whereby the court completely rejected the whole case, pronouncing completely in favour of AGAP. The plaintiffs decided not to appeal this clear legal decision. The final judgement upholds AGAP's new Constitution, AGAP's constitutional right to conduct training, now realized in ISAPZURICH, and the propriety of all Executive Committee procedures leading up to and connected with the votes on these items.

Meanwhile, the OC of ISAP tries to insure that ISAP's communications and procedures are correct – and otherwise takes precautions to avoid unnecessary provocation. For a number of committee members these efforts have sometimes meant redoubled work and constrained personal initiative. Special thanks are due to these colleagues who bear with the frustrations that result of such legal maneuvering.

Stacy Wirth, Vice President

CONTRIBUTORS

Kathrin Asper, Ph.D., was born in 1941 in Zurich. Studies in literature and education led to her doctorate in literature. Since completing training at the C.G. Jung Institute Zurich in 1975 she has maintained a private practice and lectured worldwide. Her publications in English include *The Inner Child in Dreams* (1992) and *The Abandoned Child Within: On Losing and Regaining Self-Worth* (1993). Her professional interests encompass narcissism and self-worth, trauma connected with physical disability, and psychotherapeutic perspectives on fairy tales, literature, and art.

Nathalie Baratoff, lic. phil., originally majored in Russian area studies with degrees from Brown University (USA) and Zurich University. She trained at the C.G. Jung Institute Zurich, graduating in 1987. She has been a training and supervising analyst at ISAPZURICH, as well as a regular lecturer, emphasizing fundamentals of Jungian psychology, dreams and fairy tales. She maintains a private practice in Zurich. Presently, she is ISAP's Director of Programs and author of *Oblomov: A Jungian Approach, A Literary Image of the Mother Complex* (European University Studies, 1990), and "The Golden Fish," an article in *Symbolic Life 2009*, Spring Vol. 82: A Journal of Archetype and Culture.

Paul Brutsche, Ph.D., was born in 1943 in Basel. He studied philosophy and psychology and received his doctorate in philosophy from the University of Zürich. He trained at the C.G. Jung Institute, Zürich, and has been in private practice since 1975. He served terms as president of the Swiss Society of Analytical Psychology (SGAP), the C.G. Jung Institute Zürich, and ISAP-ZURICH. He has lectured mainly on art, creativity and picture interpretation in Switzerland and different parts of the world. He is also the author of several articles in Jungian journals. In 2007 at ISAP's premiere of *The Jung-White Letters* he played C.G. Jung and has continued in this role at other venues. He has embodied C.G. Jung also in performances *of Scenes from the Red Book*, which have been presented in countries of Europe and Asia.

Irene Berkenbusch-Erbe, Ph.D., studied German literature, theology, and psychology at the Universities of Göttingen, Tübingen and Heidelberg. Until 2010 she taught in German high schools (Gymnasium). In 1996 she began her studies at the C.G. Jung Institute Zürich and received her diploma in Analytical Psychology in 2004. She is now in private practice as a psychotherapist in Ludwigshafen am Rhein. Since 2006, she has been a member of ISAPZURICH where she teaches and is a training analyst. Her areas of special interest are: dreams, trauma, pictures, religion and psychology. She has written for several publications. Since 2009, she has contributed to the IAAP training program in Poland and Lithuania.

Stefan Boëthius holds Master's & Ph.D. degrees in business and a diploma in Analytical Psychology, sits on the executive boards of various corporations that provide services to improve mental health in organizations. He has served in a honorary capacity on the Council and as treasurer of ISAPZURICH since 2004. His major assignment is with ICAS Switzerland, an international provider of employee assistance programs (EAP), where he is the major shareholder and president of the executive board. One of his vocations is to support personal growth of employees and leaders.

Katharina Casanova holds a lic. phil. degree in psychology from the University of Zürich. Since 1990 she has participated in a group practice working with adolescents, adults, and couples. She graduated from the C.G. Jung Institute Zürich in 2001. From 2004–2008 she served as Director of Studies at ISAPZURICH where she is also a training analyst. She is especially interested in the psychology of dreams, picture interpretation, and the feminist history of religion. Her article, "The Wild Feminine: Reconnection to a

Powerful Archetypal Image," was published in *Symbolic Life 2009*, Spring Vol. 82: A Journal of Archetype and Culture.

Marco Della Chiesa, Prof. lic. phil., is a sociologist and graduate of the C.G. Jung Institute Zürich. He graduated from the Moreno Institute in Stuttgart with the title "Director of Psychodrama" and is Professor Emeritus at the Polytechnic Institute of Northwest Switzerland. From 1998–2004 he served as president of the Swiss Society for Analytical Psychology (SGAP). He is a faculty member of ISAPZURICH where he was elected co-president in 2012. He lives with his family in Mönchaltdorf and conducts his private practice in Zürich.

Deborah Egger-Biniores, M.S.W., received degrees in religion and psychology from Hendrix College and a clinical degree in social work from the University of Arkansas. She graduated from the C.G. Jung Institute Zürich in 1990. She is a training analyst and supervisor for ISAPZURICH and maintains a private practice in Stäfa, Switzerland. Her professional areas of interest include developmental psychology, transference in clinical work, and the role of archetypes in individual development. Deborah has served since 2012 as a co-chair of the Jungian Odyssey.

Allan Guggenbühl, Prof. Ph.D., received his degrees from the University of Zürich in education and psychology and, afterward, his diploma from the C.G. Jung Institute Zürich in 1994. He is Director of the Institute for Conflict Management in Bern and well-known for his methods of mythodrama and crisis intervention in various Swiss schools. He has written many publications, including his celebrated book, *Men, Power and Myths: The Quest for Male Identity* (Continuum, 1997).

Judith Harris, Ph.D. is a training analyst and supervisor at ISAPZURICH. For many years she has worked extensively with Marion Woodman in BodySoul Workshops. She a pianist, a yoga teacher, the current president of the Philemon Foundation, and a board member of the Foundation for Jungian

International Training Zürich (JITZ). Her publications include *Jung and Yoga: The Psyche-Body Connection* (Inner City Books, 2000).

John Hill, M.A., received degrees in philosophy from the University of Dublin and the Catholic University of America. He trained at the C.G. Jung Institute Zürich, has practiced as a Jungian analyst since 1973, and is a training analyst and supervisor at ISAPZURICH. John served as the Academic Chair of the Jungian Odyssey from 2005 to 2012. John's publications include articles on the Association Experiment, Celtic myth, James Joyce, dreams and Christian mysticism. He recently published his book based on his lectures given for the 2010 Zurich Lecture Series in Analytical Psychology is *At Home in the World: Sounds and Symmetries of Belonging* (Spring Journal Books, 2011).

Lucienne Marguerat, lic. phil., was born in 1943 in Lausanne. She received her degree in sociology from the University of Geneva. She worked for over 10 years as a computer specialist in Zürich and then started her training at the C.G. Jung Institute Zürich. She has a private practice in Zürich and is a training analyst, supervisor, and chair of the Promotions Committee at ISAPZURICH. She has given lectures and workshops at the Antenne Romande in Lausanne, the C.G. Jung Institute Zürich, and at ISAPZURICH. Her areas of interest include fairy tales, dreams, time, the archetypal feminine and outsider art. Among her publications are, "The Importance of Kissing: The Embrace in the Crayon Drawings of Aloïse," in *Love, Traversing Its Peaks and Valleys* (Spring Journal Books, 2012) and "Balancing between Two Cultures: An Uneasy Swiss Posture," in *Unwrapping Swiss Culture* (Spring Journal Books, 2011).

Isabelle Meier, Ph.D., is a graduate of the C.G. Jung Institute Zürich (1996) and maintains a private practice in Zürich. She has further trained as a guided affective imagery (GAI) therapist. She received her doctorate in psychology from the University of Zurich and degrees in history and philosophy. She is a training analyst and supervisor of ISAPZURICH, and was elected co-president (with Marco Della Chiesa) in 2012. She served as Co-Chair of the Jungian Odyssey Committee from 2006 to 2012. Isabelle has authored several publications, such as "The Swiss as Hobbits, Gnomes, and Tricksters of Europe" in *Unwrapping Swiss Culture* (Spring Journal Books, 2011), co-editor of *Seele und Forschung* (Karger Verlag, 2006), and is on the editorial staff for the German edition of the Journal of Analytical Psychology. Her special area of interest lies in the links of imagination, complexes and archetypes.

Bernhard Sartorius, lic. theol., received his degree in theology from Geneva University in 1965 and worked for several years as a protestant minister, first in a parish and then with youth. He graduated from the C.G. Jung institute Zurich in 1974, maintaining his private analytical practice first in Geneva and, since 1997, in Lucerne and Zürich. He is a training analyst and supervisor at ISAPZURICH. Among his publications on symbolical subjects are the essays, "Eros and Psyche Revisited," in *Love: Traversing Its Peaks and Valleys* (Spring Journal Book, 2013), and "La Mecque ou/où on meurt," in *Vouivre, Cahiers de psychologie analytique, Pèlerinages* (Numéro 11, 2011), and his book on the Orthodox Church, *L'Eglise orthodoxe, Grandes religions du monde*, Vol. 10 (Edito-Service, 1982).

Kristina Schellinski, M.A., holds a degree in political science and literature. From 1983 to 1998 she worked for the United Nations Children's Fund (UNICEF). She graduated from the C.G Jung Institute Zürich in 2002, culminating training with her thesis, "Oh, Brother – A Woman's Search for the Missing Masculine: A Jungian Perspective on the Challenges and Opportunities Faced by a Replacement Child." She has taught at the C.G. Jung Institute Zürich as well as at ISAPZURICH and has lectured at international congresses. She is a founding member of the Rencontres Jungiennes at Lavigny, Switzerland.

Murray Stein, Ph.D., Canadian born, completed his university education in religion and psychology in the U.S.A. and trained at C.G. Jung Institute Zürich. Today he is a training analyst and supervisor at ISAPZURICH where he previously served as president. He is a former president of the International Association for Analytical Psychology (IAAP) and a founding member of two IAAP societies: Inter-Regional Society for Jungian Analysts (USA) and the Chicago Society of Jungian Analysts. He has authored many books, including *Jung's Treatment of Christianity* (Chiron, 1985) and *Minding the Self: Jungian Meditations on Contemporary Spirituality* (Routledge, 2014). He is the editor of *Jungian Psychoanalysis: Working in the Spirit of C.G. Jung* (Open Court, 2010). With Nancy Cater he is co-editor of the *Zurich Lecture Series in Analytical Psychology* (Spring Journal Books), which is a compilation of the weekend of lectures co-hosted every autumn by Spring Journal Books and ISAPZURICH.

Ilsabe (Bille) von Uslar, lic. phil., holds a degree in psychology from the University of Zürich. She participates in a group practice working with children, adolescents, and adults. She has given lectures and workshops at both the C.G. Jung Institute Zürich and ISAPZURICH. Her areas of interest include dreams, picture interpretation, imagination, and techniques of relaxation and trance induction. She is married and has four adolescent children.

Stacy Wirth, M.A., received her B.A. in dance and anthropology from Mills College and her M.A. in the psychology of art from Antioch University. In 2003 she graduated from the C.G. Jung Institute Zürich and since then has conducted a private analytic practice in Zürich. Elected to the AGAP Executive Committee in August 2004, she served as secretary until 2010 and as co-president from 2010 to 2013. She also served from 2004 to 2010 as a member of the ISAPZURICH Council, first as the vice president and later as secretary. Stacy is currently a training analyst at ISAP and Co-Chair of the Jungian Odyssey Committee.

Ursula Wirtz, Ph.D., is a Jungian training analyst and graduate of the C.G. Jung Institute Zürich (1982), maintaining a private analytical practice in Zürich. She received her doctorate in philosophy from the University of Munich and her degree in clinical and anthropological psychology from the University of Zürich. She is a faculty member of ISAPZURICH, a trainer with developing Jungian groups in Eastern Europe, and has taught at a number of European universities. Ursula has authored many publications on trauma, ethics and spirituality, for instance, *Seelenmord: Inzest und Therapie* (Kreuz, 2005). Her book, *Trauma and Beyond: The Mysteries of Transformation* (Spring Journal Books, 2014) ensues from her presentation at the 2012 Zürich Lecture Series in Analytical Psychology. Ursula succeeded John Hill as Academic Chair of ISAPZURICH in 2012.